ELECTRICIAN'S
Technical Reference

Variable-Frequency Drives

ELECTRICIAN'S
Technical Reference

Variable-Frequency Drives

Robert S. Carrow

DELMAR

THOMSON LEARNING ™

Australia Canada Mexico Singapore Spain United Kingdom United States

DELMAR

THOMSON LEARNING

Electrician's Technical Reference: Variable-Frequency Drives
Robert S. Carrow

Business Unit Director:
Alar Elken

Acquisitions Editor:
Mark Huth

Development:
Dawn Daugherty

Executive Marketing Manage:
Maura Theriault

Channel Manager:
Mona Caron

Executive Production Manager:
Mary Ellen Black

Project Editor:
Barbara Diaz

Art/Design Coordinator:
Rachel Baker

Library of Congress Cataloging-in-Publication Data

Carrow, Robert S.
Electrician's technical reference. Variable frequency drives / Robert S. Carrow.
p. cm.
Includes index.
ISBN 0-7668-1923-X
1. Electric motors, Alternating current. 2. Electric driving, Variable speed. 3. Variable speed drives—Automatic control. I. Title.

TK2781 .C327 2000
621.46—dc21   00-030690

Notice to the Reader

Contents

Dedication

To my parents Robert O. and Ramona J. Carrow, who gave me God's gifts,
and
To my wife Colette, my son Ian, and my daughter Justine, whose patience helped to make this project a success.

Preface

As industry and commerce grow, the equipment used for that growth changes. These technological changes allow for more production, faster speeds, and better products with less waste. With the technological change comes the need for complete understanding of this new equipment. This new, electronic equipment is computer-based and built for efficiency, and most is solid state in construction. Variable-frequency drives (VFDs) fall into this category of new, electronic equipment that is changing. Understanding how they are constructed, how they work, and what to do when they fail will help industry and commerce to continue with phenomenal growth.

The individuals responsible for the selection, application, and maintenance of these VFDs are many. They are electricians, mechanics, HVAC technicians, engineers, designers, programmers, and even managers. They all need a good understanding of VFDs. This book is written for them. It contains all the relevant information concerning variable-speed systems, which are made up of the electric, mechanical, and computer disciplines. The book is also written from more of a practical vantage point than theory. Basic theory is presented whenever necessary.

Chapter 1 covers variable-speed basics. Traditional variable-speed systems are discussed as well as the topic of speed regulation and motion control. Here, the reader is also introduced to solid-state controllers. Chapter 2 builds on the base from the first chapter. To understand how the drives work and interact with their loads, one must review the concepts and relationships of horsepower, speed, and torque. Inertia is discussed along with the mechanics of a drive system. From here, electric basics are covered and how they relate to VFD installations. They include power factor, capacitors, inductors, frequency, and power.

Once the basics are covered, the book then goes into AC-induction-motor design and construction. To understand the motor controller (the VFD), one should know as much as possible about the motor being controlled. Chapter 3's remaining text covers all types of VFDs, their construction, theory of operation, inverter-duty motors, PWM concepts, VVI, VSI, CSI, flux-vector control, direct-torque control, and field-oriented control. Chapter 3 contains the most information of any chapter in the book.

Chapters 4, 5, and 6 dwell on issues concerning the installation, troubleshooting, and metering of the VFDs. Voltage spikes, drive location, heat dissipation, carrier frequency, harmonic distortion, fault handling, diagnostics, maintenance, waveforms, testing, and proper setup are all covered in these chapters. Several troubleshooting trees for common VFD faults are also provided. Chapters 7 and 8 deal with the sizing, selection, and application of drives in commercial and industrial installations. Torque and load conditions, process control, automation, and ambient conditions are all discussed.

The last chapters deal with the energy-saving ability of the VFDs, how to identify those opportunities, discussion by the U.S. Department of Energy regarding VFDs and motors, and how to make any facility more efficient (thus saving valuable energy dollars). New VFD technology is addressed along with future applications within the residence and with electric vehicles. Useful conversions and formulas are part of the appendix, and there is an extensive list of acronyms used in the industry.

This book *will* help keep facilities in operation and keep their motors running. It will pay for itself by identifying problem areas regarding VFDs and motors or showing how the VFD, when applied, will save money or make a process better.

The author and Delmar wish to acknowledge and thank the people who have provided reviews and contributions to this book. Thanks go to:

E. David Brown, Sr.
The Boeing Corp.
Marysville, WA

Jim Boyd
National Joint Apprentice Training Committee
Upper Marlboro, MD

Kevin Early
Waterford, PA

CHAPTER

1

Variable-Speed-Drive Basics

Speed and Motion Control

If some machine or process is moving, if air is being blown, or if water is being pumped, then there is *motion*. Motion can be caused by mechanical, pneumatic, hydraulic, or even electric means. There are also several technologies that have been employed that provide the necessary control of the process or machine. Some are old, some are inefficient, and some are expensive. With motion comes the need for *motion control*. Motion control can be as simple as opening and closing a valve to control how much water flows through a pipe, or it can be as complex as programming a six-axis robot to perform multiple, circular, interpolation-based moves simultaneously. Motion control is:

- The starting or stopping of a machine or process
- The opening or closing of discharge valves in pump systems to control flow
- Using a vortex damper or inlet guide vanes to control airflow
- The control of any prime mover
- The speed, torque, or position control of an electric motor

All motion requires some type of *prime mover*. Prime movers can be electric motors, diesel engines, air, or even hydraulic systems. The starting and stopping of electric motors is motion control; so too is the air-actuated plunger knocking bad parts from the assembly line conveyor. Many methods and disciplines are used when it comes to motion control. Computers, electric systems, the mechanical components, and all the other peripheral devices make motion control a discipline unto itself. Motion control can also be broken down into several individual categories as already mentioned. They are speed or velocity control, torque control, current control, position control, and starting or stopping. The accuracy and performance required dictate how much control is needed over a certain category. Just to have motion, equipment is needed that can initiate that motion. This is the starting equipment.

A *starter* is just that—a device used to start a machine or process. It typically is not intended for any type of variable speed control. It can soft-start the application or it can engage motion quickly and abruptly. Obviously, a soft-start approach is much better for

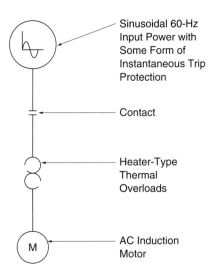

Figure 1–1   Full-voltage contact starter.

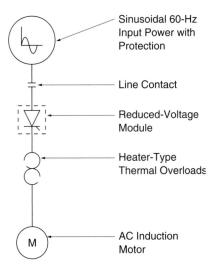

Figure 1–2   Reduced-voltage starter.

the application's mechanical components and its drivetrain. There can be electric or mechanical means of starting; yet, in either case, there has to be a prime mover. The prime mover is most often either an electric motor that is started or a motor that is already running, in which case motion is engaged mechanically.

Full-voltage-contact starters have been employed for many years. As can be seen in Figure 1–1, these starters provide a switch to allow electric current to flow to the motor and start it running full speed. They typically include thermal or electronic overload devices and instantaneous protection devices to keep the motor from drawing too much current and subsequently failing. The *reduced-voltage starter* is shown in Figure 1–2 and is

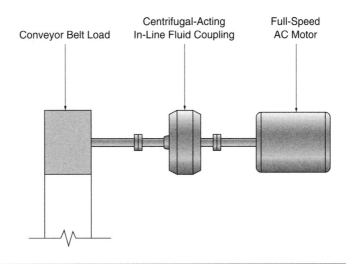

Figure 1–3 A fluid coupling that works by way of centrifugal force.

a more expensive device used to start an electric motor. It employs thyristors only on starting, to limit the voltage and current to the motor. This provides for an acceleration ramp up to the motor's full speed. The reduced-voltage starter also provides overload protection for the motor.

There are also other means of starting a machine or process that are mechanical in nature. One such means is the *fluid coupling,* Figure 1–3. A fluid coupling works in conjunction with a prime mover to softly engage or start a shaft. By using hydraulic fluids and centrifugal forces, the fluid coupling softly brings the driven component up to full speed. No variable speed control is attained. Likewise, a process may require peripheral components to be on or off, selectively. These on and off states of motion can be controlled by engaging and disengaging a mechanical clutch that is connected to a prime mover. Again, no variable speed control is attained with this equipment.

Reasons to Consider Using Variable-Speed Equipment

The previously mentioned starter equipment cannot be used to change the speed of a given machine or process. Any time some speed lower than the full, base speed of the system is required, *variable-speed drives* must be considered. Better stated, any time the speed has to change (off of full speed) and conventional gear/speed reduction will not do, then variable-speed drives must be employed. Merely slowing the speed of a process or machine down by reducing its speed with gearing still does not allow for speed changes. For example, using a 1750-rpm, full-speed motor and gear reduction of two to one yields an output speed of 875 rpm but no speeds above or below 875. This is when variable-speed drives have to be used.

Two other reasons that variable-speed drives are attractive are energy savings and the fact that they can be easily incorporated into the modern-day automated factory. An electronic AC variable-speed drive can operate an electric motor at a reduced speed and at a lower level of power. This principle is covered in detail in Chapter 3. As for the automated factory concept, electronic variable-speed-drive equipment lends itself well to being interconnected to other electronic factory devices. Microprocessors, feedback devices, and communications capabilities make variable-speed drives useful to the automated facility.

Traditional Variable-Speed Systems

As already discussed, motion in the factory, the commercial facility, and virtually all around us is mainly actuated by either pneumatics (air), hydraulics (liquid), steam, or electricity. Steam is used in power plants to drive turbines to generate electricity and by some industrial plants for internal use. Hydraulic and pneumatic systems are used to some extent in most facilities. However, there are disadvantages associated with each of those methods. Pneumatic and hydraulic systems tend to be high in maintenance. Hydraulic systems tend to leak and are usually dirty. This makes them unsuitable for industries needing clean plant environments, such as the food industry. Pneumatic systems often get water in their air lines, and they are noisy. Pressure losses equate to poor performance.

Nowadays, pneumatic and hydraulic uses in motion control are either specialized or supplemented to machines that are mainly controlled electrically. Many times, using a pneumatic or hydraulic solution is appropriate only because of initial costs, but in the long run, these kinds of systems are inefficient.

The introduction of the induction motor nearly a century ago has allowed various methods to reduce the speeds for an application. Electrically powered machines used these constant-speed AC motors as the prime mover. The DC motor evolved simultaneously with the AC motor but needed an electric rectifier. It has been known for years that AC motors that ran at their full speed were energy wasters when restricting the speed mechanically. This technology also did not allow for any closed-loop velocity or position control. Since the motor ran full speed, gear reduction had to be incorporated to run the machine at an appropriate slower speed. Mechanical solutions, such as brakes and clutches, initially gave the plant people some means of control. Alternating current motors were a necessity in the factory, even though controlling them was another issue. The need for variable-speed drives had arisen, and engineers employed the available technologies (Table 1–1).

These engineers and designers came up with various techniques for reducing the speed of an electric motor. These centered around AC motors. Many approaches were mechanically based. Some were industry and application specific, while others incorporated hydraulics as the methodology. In many cases, the AC motor still had to run at full, constant speed, thus making the device still very inefficient. Not until electronic AC drives emerged on the scene did technology start getting away from these high-maintenance, multicomponent, energy-wasters.

One such device is the *mechanical drive* or *variable-pitch pulley,* sometimes also called the V-belt drive (see Figure 1–4). Motion commences with a full-speed AC motor, which is attached to one of the pulleys. By adjusting the distance between pulleys and by changing the actual depth of the pulley that can be opened or closed, the output can be changed. The V-belt moves up and down within this adjustable groove. This speed controller is employed throughout industry and is still used today, but in dramatically reduced numbers. The reasons for its demise include limited speed range turndown, belt slippage during acceleration,

TABLE 1–1 Traditional Variable-Speed Systems

Mechanical Drive (Variable-Pitch Pulley)
Hydraulic Drive (Hydrostatic Speed Variator)
Transmission (PIV) Packages
Eddy Current Coupling (Clutch)
Rotating DC (MG Set)
Gear- and Speed-Reduction Systems

general V-belt wear and breakage, limited horsepower sizes, and impractically of soft-starting of the load. It was good in its time because it was a simple, inexpensive design.

The fluid-based, variable-speed drive is the *hydrostatic speed variator* or *hydraulic drive* (see Figure 1–5). By configuring internal jets, valves, and impellers a certain way, the fluid can flow against those parts, producing motion. The design varies from manufacturer to manufacturer. It offers soft starts, reversing, and a wide speed range. It can run at very low speeds and provide constant-output torque. However, it gives off a lot of heat, needs a periodic oil change, costs more initially, has limitations as to how high in horsepower it can be built, and needs a constant-speed AC motor on the input.

Elaborate, expensive *transmissions* offer a similar speed control scheme but without using a fluid as the basis for motion. Operating much like the automatic transmission of

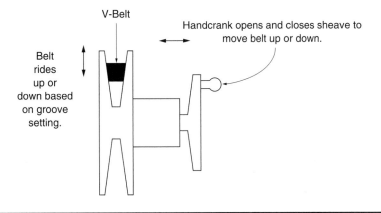

Figure 1–4 A mechanical drive or variable-pitch pulley.

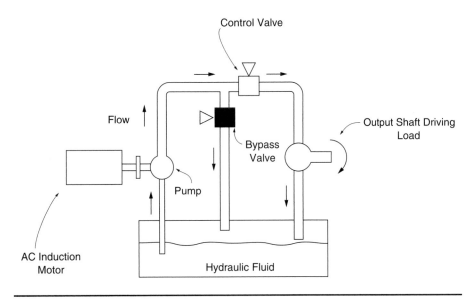

Figure 1–5 A hydraulic, variable-speed system.

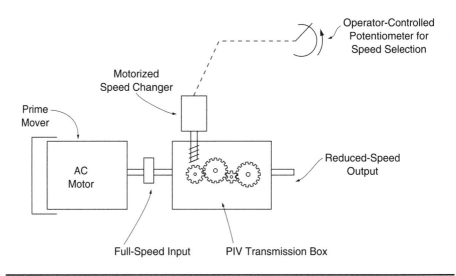

Figure 1–6 PIV or transmission case for speed variations using gears.

the common automobile, these transmission boxes, sometimes called *PIVs,* have many high-precision gears housed within. A smaller motor is often employed to move the internal gears into place based on the "dialing-in" of a requested speed (see Figure 1–6). The machining of these gears and their sheer quantity make these devices very expensive. The speed range is limited, and, if a gear breaks, other mating gears may break, and they cost a fortune to repair. They need to be lubricated and maintained and are not very efficient. An AC motor also must provide motion as the prime mover. Additionally, there is no real soft-start capability and reversing can only be achieved after the transmission is stopped.

An *eddy current coupling* or *clutch* is shown in Figure 1–7. As the name implies, the eddy current principle is in use here. Eddy currents, induced within the conducting material by the varying electric field, cause the desired effect of changing output speed, due to the slip between the constant-speed shaft and the variable-speed shaft. Soft starting and high torque output, especially at low speeds, make this device a workhorse. The main disadvantage is that the clutch system is very inefficient and must be cooled, either by water or air, and there is still a lot of energy lost as heat. The eddy current clutch was, in its heyday, probably the most widely used variable-speed solution. However, it still needed a constant-speed AC motor as the prime mover.

The *rotating DC* or *motor-generator (MG) set* was an initial method of controlling the speed of the DC motor. It was actually comprised of an AC motor and a DC generator (see Figure 1–8). The AC motor was coupled to a generator to produce DC, which could be controlled to change the speed of a DC motor. The MG set's prime mover could also be gasoline- or diesel-powered to run the generator, thus still eventually powering the DC motor. This was ideal for remote and isolated locations. The MG set was the mainstay in industry for years, providing DC power to DC motors. With new AC and DC control technology, MG sets are no longer very popular. They are very inefficient and use more energy than is needed to perform the job. It is also hard (if not impossible) to find replacement parts. Out of the MG set evolved the static-controller scheme for DC motors.

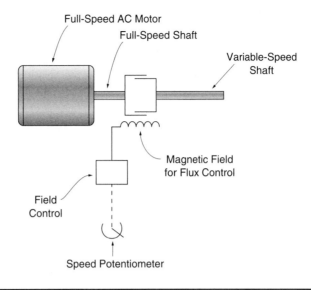

Figure 1–7 Eddy current coupling or clutch.

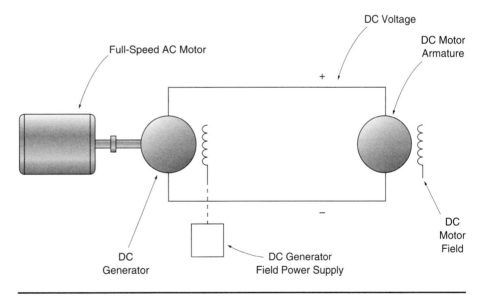

Figure 1–8 Motor-generator set (MG set).

Another method of achieving the desired speed output is with *gear reduction* or *speed reduction*. This is an inexpensive and straightforward way to get a needed speed, but speed changes are not easily made. As can be seen in Figure 1–9, a prime mover generates a constant speed, usually at a high rate. The gear system is in place not only to reduce that speed but also to provide an increase in torque. As the prime mover runs, the output of the

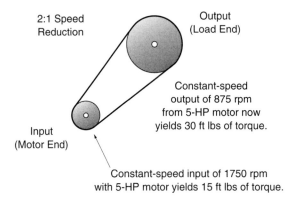

Figure 1–9   Gear reduction provides speed reduction.

system runs at some constant slower speed. To make a speed change, the gear-reduction ratio must physically be changed.

Static and Solid-State Variable-Speed Controllers

As can be seen with most of the previous variable-speed examples, speed changes are *dynamically* achieved. There is a prime mover running at some higher constant rate of speed, and speed changes are effected mechanically from that prime mover. This is a very inefficient way to utilize power. Each dynamic system has a dramatic loss attributed to it due to friction, noise, or heat. A more efficient use of power and energy, especially whenever electricity can be used, is that of *static* or *solid-state controllers*. Reasons to use static variable-speed-control systems include:

- Improved and more efficient control scheme
- Energy savings
- Improved diagnostics capability
- Network and serial communications
- Improved mechanical characteristics
- Bypass for AC drives

Solid-state or static devices were initially developed mainly for DC motors to achieve variable speed control. Static devices could be equated to any power-conversion system that did not require a motor driving a generator to achieve the desired electric output. In other words, nothing was used mechanically or dynamically to convert AC to DC or DC to AC. The power conversion would be handled through early, strictly electric, means. Power-converting devices took in AC electric energy in one form, and DC energy emerged in another form. Vacuum tubes (or electron or thyrotron tubes) were used to change the DC voltage to the DC motor's field or armature. These systems took up a lot of space on the factory floor and also gave off a lot of heat. Today, it is difficult to even purchase replacement parts for these types of systems. Many are being replaced with digital DC drives or AC drives.

$$\% \text{ Drive Efficiency} = (\text{Power 2 / Power 1}) \times 100$$
$$\% \text{ Motor Efficiency} = (\text{Power 3 / Power 2}) \times 100$$
$$\% \text{ System Efficiency} = (\text{Power 3 / Power 1}) \times 100$$

Figure 1–10   Drive, motor, and system efficiencies.

TABLE 1–2   A Comparison of System Efficiencies

Drive Type	100% Speed	75% Speed
Variable-Pitch Pulley	.82	.72
MG Set (DC)	.65	.61
Eddy Current Clutch	.86	.64
Solid-State Drive	.87-.89	.85

Variable-Speed-Drive-System Efficiency

Throughout this chapter, various drives have been discussed. There are advantages of one type over another and vice versa. Some variable-drive systems are more efficient than others. This ultimately causes some systems to provide better speed regulation or performance than others. More importantly, in consuming power wisely, variable-speed-control systems have to be analyzed based on their efficient (or inefficient) use of energy. The efficiency of any device is expressed as the ratio between the power output and the power consumed. In the case of the solid-state drive in Figure 1–10, the difference between the input power and the output power is the power loss of the drive, which is dissipated as heat. The efficiency can be calculated for the solid-state variable-speed drive alone, the motor alone, or for the drive and motor system. If a transformer were included, then its efficiency could be calculated and it could also be factored into the system's overall losses.

Using Figure 1–10 as a basis, other variable-speed systems can be analyzed for their own inherent efficiency. In the case of the variable-pitch pulley, there are losses with a prime mover running full speed and full load and the friction of the belt riding in the sheaves. This method of speed control coincides with losses. The same is true for the MG set and the eddy current clutch system. These system efficiencies are compared in Table 1–2.

From the values shown in Table 1–2, it is apparent that those systems dependent on dynamics rather than statics (the use of an electric motor as in the eddy current system or the MG set) were much less efficient. The presence of an electric motor has the most significant influence on a system's efficiency. The motor draws electricity, and at constant load this power is higher than the mechanical motor output to the shaft. This is due to losses within the motor. These losses are mainly copper and iron losses. The copper loss is a result of the resistance in the stator and the rotor, whereas the iron losses are comprised of the eddy current and hysteresis losses. These hysteresis losses occur whenever the iron is magnetized by the AC. As the motor is magnetized and demagnetized repeatedly, energy is required. As the frequency increases, the losses increase as the eddy currents generate heat within the core. The motor core is composed nowadays of thin, insulated sheets to reduce the cross-sectional area where the eddy currents flow and to break the flow, thus reducing the heat loss. Ventilation losses also occur due to the air resistance of the fan, while friction losses are present due to the wear on the ball bearings. Thus, there are several forces at work within the common electric motor that contribute to its power losses. All told, losses in any device can directly affect its performance. When variable-speed drives are used, performance is typically equated to speed regulation, that is, the ability to run the machine or process at the desired speed.

Speed Regulation

Whenever good speed regulation is the requirement, merely changing the speed may not be adequate. The machine or process has to be able to reach a desired speed and maintain the speed even while under severe or changing load conditions. Thus, there will always be some factor or value of speed regulation attributed to any variable-speed-drive system. Motion control, as it pertains to controlling electric motors, also implies *regulation*. This regulation is the drive controller's function. Speed and current regulation make or break the application as far as performance is concerned. How much regulation is required will also determine what the initial costs will be for drive equipment. For varying degrees of speed regulation, there will be a direct correlation to cost. A more elaborate and expensive system typically ensures better speed regulation.

Speed regulation implies that an expected percentage, or actual motor revolutions per minute (rpm), can be measured at the motor. If speed regulation is 3 percent for a 1750-rpm motor, then this means that the speed can be off by 52.5 motor rpms. Can the application tolerate this? If not, then a better, more expensive approach may be in order. What is usually ambiguous is on what conditions this speed regulation is based. The loads obviously have the main impact. Are the loads steady-state, changing rapidly, severe, or light? These conditions have to be considered and compared with the predicted speed regulation to achieve optimum system performance in the application.

Speed or velocity control sounds simple but is not. Take a motor, drive, and some electricity. Now, run that motor at half speed. Accomplishable, yes, but what really happened to get the desired result? First, someone had to physically mount the motor and drive somewhere. The motor has to be mechanically connected to the load or machine it is driving (hopefully, there are no other mechanical or electric problems in the machine). Then, all of the proper wires had to be pulled to the correct terminals at motor and drive. Then, means of control had to be considered, which meant that more wire had to be pulled. At this point, it is at least ready for application of power. The motor should be uncoupled from the load, and it should be run to see if shaft rotation and direction are correct. It should also be checked, whenever possible, for commanded speed with actual speed. If there is a tachometer, a speed feedback device, is it connected properly and scaled accordingly within the drive? Checking to see that the tach feedback is "in sync" is also necessary.

As discussed in Chapter 2, driving the load or machine is actually doing the work. Work means production, but to do work the motor has to produce torque. Torque is needed to move the motor's shaft and to move that which is attached to the motor's shaft. An electric motor demands more electric current to produce more torque. Again, stressing the importance of constantly driving the load at a given speed, the electronic drive and motor give torque and current priority over speed. This means that, in a particular process, a motor's actual rpms "sag" so that there is enough usable current to continue producing torque to drive the given load. Many electric motors continue to seek more current to drive a load. This is more true with DC motors than with AC motors, but an electronic drive ahead of the motor ensures that the motor does not damage itself by requesting too much current. Torque, speed, and power are very important to an application's success, and they are all interrelated. The next chapter explores these power-transmission basics in detail.

As has been seen throughout this initial chapter, there are many methods and many systems for changing the speed of a machine or process. Depending on the application and the needs of that application, speed-changing equipment is selected. Unfortunately, once the selection is made and valuable dollars spent, it is very difficult to change or correct for situations where speeds cannot be attained or held. When designing a speed-control system, flexibility is needed with the equipment chosen. This makes solid-state, electronic, variable-speed drives attractive. They can provide good efficiencies along with energy savings and the required speed changes. However, as has been shown, there were a great many methods and systems employed before solid-state systems to perform variable-speed applications.

CHAPTER 2

Electric and Power-
Transmission Basics

Electricity and Power Transmission

The conversion of electric energy into mechanical power is actually *power transmission,* sometimes called *PT* (not to be confused with the acronym PT, standing for potential transformer). Electric energy somehow has to provide motion in a machine or process. Most often, this is first done by energizing an electric AC or DC motor. Attached to the motor shaft are other components (couplings, gear boxes, belts, clutches, brakes, gears, rollers, etc.), which further contribute to the full-power transmission to the actual load being driven. Volts change to watts and end up as torque, moving products "out the door." Much of the mechanical, power-transmission equipment now has some electric control. The plant personnel must have a good understanding of what is taking place mechanically to assimilate the electric needs. This is particularly true when applying and troubleshooting variable-frequency drives (VFDs). Plant personnel have to understand the mechanical aspects of what it is they are trying to move to keep the electronics working properly.

No matter which type of VFD is used (volts/hertz, flux vector, etc.), the net result is that a desired speed must be attained and enough torque has to be produced to continue driving the load at that desired speed. Speed regulation, as discussed earlier, is the VFD's ability to get to or remain at a desired speed under all load conditions. Some VFDs are better than others at accomplishing this, and there are several methods available to obtain better speed regulation. Applications should dictate how tight speed regulation should be and which methods are employed to get the machine or process speed correct. *All VFDs must:*

- *Run a machine or process at a desired speed*
- *Produce adequate torque to handle the load*
- *Use power efficiently to produce the necessary torque at a given speed*
- *Effectively monitor the application or process*

Speed, torque, and *horsepower* are all elements of motion control that are very much interrelated; this is evidenced by the formula in Figure 2–1. When a change is made to one of the elements, another variable is changed. The interrelationships of this formula can be seen in the speed-torque-horsepower nomograph in Figure 2–2. This gives an excellent

$$HP = \frac{T \times rpm}{5250} \quad \text{and } T = \frac{HP \times 5250}{rpm}$$

where

$$\text{Horsepower (HP)} = \frac{(V \times I \times \text{Efficiency})}{746} \quad \text{and } \text{Speed (rpm or N)} = \frac{120 \times Hz}{\text{Number of Motor Poles}}$$

Hint: To easily remember this formula, you may assume that 1 HP equals 3 ft lbs of torque at 1750 rpm (a common motor speed).

$$HP = \frac{T \times rpm}{5250} \quad \text{or, using the example, } \quad 1\ HP = \frac{3 \times 1750}{5250}$$

Figure 2–1 Speed-torque-horsepower formula.

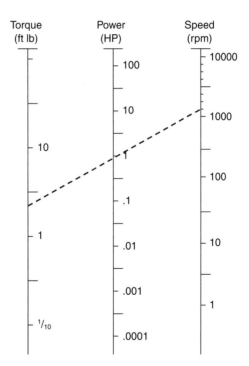

Example: At 1750 rpm, a 1-HP system yields 3 ft lbs of torque.

Figure 2–2 Speed-torque-horsepower nomograph.

graphic comparison of the three values and how they relate. It is laid out in such a way that a point, such as speed, may be selected on the graph. This can be considered one of the *known values*. If either torque or horsepower is also known, then the corresponding line through the two points (the *known components*) can be drawn and the unknown value found. This chart can save someone the time needed to calculate the values using the equations previously mentioned.

Horsepower

Sometime during the late eighteenth century, James Watt of Scotland determined that one horsepower was equal to 33,000 foot pounds of work in one minute. This is equivalent to the amount of power required to lift 33,000 pounds one foot in one minute. These values were attributed to the comparable amount of work a horse could do (see Figure 2–3). Thus, the term *horsepower* was applied and it stuck. Horsepower can be derived from the following equation, where F is force, D is distance, and t is time:

$$Horsepower = \frac{F \times D}{33,000 \times t}$$

Selecting and sizing motors strictly on horsepower has probably caused more costly undersizing and excessive troubleshooting than anyone would want to admit. The fact of the matter is that, when sizing, the torque requirements of a given application are the most critical. For motors and drives, it is important to then look at the electrical current requirements of the application and how they relate to torque. The electric equivalent of one horsepower is 746 watts. Variable-frequency drives and motors are rated both in horsepower and

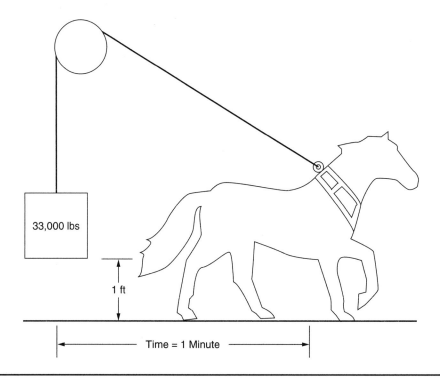

Figure 2–3 The concept known as horsepower.

in current, actually full-load current, or full-load amperes (FLA). To be more exact, when sizing the actual power requirements and selecting a motor, *brake horsepower* (BHP) is the term used. Many VFDs are rated on their continuous current output and not on horsepower. In some applications, peak, accelerating, or starting torques and currents are more important to the sizing. The actual power requirements begin at the motor shaft, with the load. It is always good practice to check the torque requirements of the driven load and make sure the motor can produce the necessary output. *A motor has a maximum amount of torque that it can produce!* Motor speed versus torque curves is available to help in this necessary check.

Torque

The term for a rotating force is *torque,* which is any twisting, turning action requiring force. As is shown in Figure 2–4, torque is the product of some force times the radius, or $T = F \times r$. If force F is applied to the lever arm at a distance equal to the radius r (shown), then a resulting torque is produced. Derivations of this formula can give the amount of work as a torque acting through an angular displacement. There are mainly two types of torque, *static* and *dynamic.* Most often, factory personnel are concerned with dynamic torque. A rotating apparatus exhibits dynamic torque. It is in a state of constant movement, correction, and change. Static torque, on the other hand, is a more consistent value. Because many prime movers exhibit rotational movement, the torque has to be factored as the work necessary to cause the angular displacement of the rotating element, as is seen in Figure 2–5.

Torque, and thus load current, is definitely the more appropriate value needed to determine what size prime mover is required to move the load accordingly. If the prime mover is an electric motor, then torque can be directly proportional to current in the rotor, as the magnetic field is considered constant. Typically, with 4-pole motors, the magnetizing current component is 25 percent of full load and the other 75 percent is the actual torque-producing current. Once the torque value is known, then a determination of which motor to use is much closer. The problem is that motors are sized by horsepower, which can be potentially dangerous. The better scenario is to provide the motor supplier with speed and torque requirements at those speeds so as to size the motor for duty cycle and complete heat dissipation. Let the motor supplier, who is knowledgeable about motor constructions, select the motor and frame. Some particular frame sizes of motors may be capable of more in the way of overloading than others.

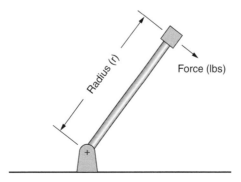

Figure 2–4 Torque equals force times the radius.

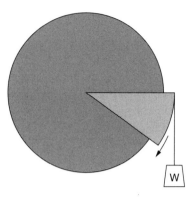

Work = Torque × Angular Displacement

$$HP = \frac{T \times rpm}{5250}$$

$$T = \frac{HP \times 5250}{rpm}$$

Figure 2–5 Torque produced about a motor's shaft.

When in doubt as to the torque requirements of an application, physical measurements can be made to determine load requirements. A *torque wrench* is sometimes used to tighten nuts onto a seated surface and can be employed to measure loads. This device typically measures the amount of force in *in lbs* (inch pounds). A similar device, a *torque meter,* can also be used to measure the torque requirements of a particular shaft. This evaluation is good for any rotating component but will not factor into the equation any acceleration or peak torque requirements. A motor is a stupid device, and, if it cannot turn because its load is too great, then the current to the motor will increase to try and move that load until the supply is shut off. That is why instantaneous and thermal, electronic, overload-protection devices should always be implemented into a motor system, even when using a VFD (*National Electrical Code*® Article 430 requires thermal overloads (TOLs) with a VFD applied to a motor). When the AC motor will not run, it may be undersized and the torque output from that motor inadequate to perform the application.

To start a loaded AC motor, it may be more important to consider *breakaway* or *starting torque* than all other values of torque in the operating cycle for that machine or process. *Running torque* is typically the first concern, and it should be, as the application may be sensitive to torque variations or load changes once it is at rated speed. Running torque can be expressed, too, as simply the torque required to just keep the machine running at a given speed. *Process torque* could factor into the requirements the need to compress, cut, or act on a material periodically, possibly increasing torque needs for an instant. Lastly, the *acceleration torque* needs of an application have to be analyzed. All too often, someone wants a machine or process to cycle faster; thus, this torque value plays the predominant role in sizing, especially if there is a high-inertia component in the system. Nuisance overcurrent trips may occur.

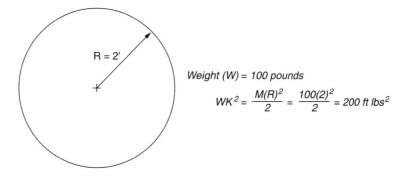

$R = 2'$

Weight (W) = 100 pounds

$$WK^2 = \frac{M(R)^2}{2} = \frac{100(2)^2}{2} = 200 \ ft \ lbs^2$$

$$Acceleration \ Torque \ (T_a) = \frac{WK^2 \times rpm}{(308 \times Time)}$$

Given: Acceleration is 600 rpm in 10 seconds

Therefore: $T_a = \dfrac{200 \times 600}{(308 \times 10)} = 38.96 \ ft \ lbs \ Required$

And: $HP = \dfrac{38.96 \times 1750}{5250} = 13 \ HP$, so use a 15-HP motor.

However, if acceleration is to be twice as fast, say 5 seconds, then the torque required is almost 78 ft lbs and the motor size will double. Gear reduction can help lighten the load reflected back to the motor.

Figure 2–6   Inertia and torque due to acceleration.

Inertia and Acceleration Torque

Beyond the linear and rotary torque and horsepower basic formulas, there are several other useful equations to aid in sizing motors and prime movers. One is used to find to the acceleration torque of a rotary device. *Acceleration torque* is the torque required above and beyond the torque required to drive the given load. It must be added to the usually larger value of load torque to adequately size an application with substantial acceleration rates. This is found by the formula:

$$Acceleration \ torque \ (T_a) = \frac{(WK^2) \times change \ in \ speed}{308 \times t}$$

where:

T is expressed in foot pounds (ft lbs), and
(WK^2) or sometimes expressed as J, (joule), is the total system inertia, which includes motor inertia (from the rotor) and the load's inertia and is expressed in foot pounds squared.

If a gear reducer or other important inertia-containing component is part of the system, then its inertia value must be included also. Any power-transmission component that must be part of this acceleration has to be factored into the equation. *Inertia,* that property whereby an object in motion wants to stay in motion and that same object at rest wants to remain at rest, is many times the overriding factor in sizing prime movers. Inertia is seen graphically in Figure 2–6, and an example is given. The change in speed, in rpm, is sometimes shown as N.

From what current speed to what new speed? Zero to 1750 rpm? Half speed to three-quarter speed? This must be factored into the equation. Then, a desired time to accelerate this load to desired speed in seconds must be selected. If the high-inertia load cannot be accelerated in the desired time, then the time to get to speed must be lengthened or the effects of the inertia reflected to the motor must be "relieved."

Electric Basics

Thus far, power transmission has been observed from primarily a mechanical vantage point. Ahead of this conversion—electric to mechanical—are millions of electrons just waiting to get involved. In drive and motor systems, electricity and magnetism are the basic elements behind the scenes. They were once thought to be two separate forces, but Albert Einstein's theory of relativity showed that the two were *related*. Electricity is produced from basic electric charges. Magnetism is created by moving electric charges and reacting to them with other electric charges (i.e., creating a magnetic field in a motor and cutting into that field with a conductor full of current). Electromagnetic forces and magnetic flux are all a result of the presence of electricity. *Electricity* is a form of energy that consists of mutually attracted protons and electrons (positively and negatively charged particles).

Electric activities or functions may not be apparent. The raw, incoming power to a VFD for rectification, as in a solid-state device, is one level. The action is not readily noticed as it would be with an AC motor driving a DC generator that eventually provides power to a DC motor. Another level is that of control and communications power to pass data. Actual movement of a load is caused by an electric prime mover or motor generated by higher levels of electric energy and converted into mechanical energy. It all originates with the power-producing utilities and is distributed in two forms: AC (alternating current) and DC (direct current). Practically all electricity is transmitted in an AC form and is changed to DC whenever DC is needed.

From the utility to the substations outside any facility to the internal switchboards, electricity is usually distributed throughout via transformers. These devices actually provide the usable, nominal 60-Hz voltages—575 volts, 460 volts, 380 volts, 230 volts, and 115 volts being the most common. In addition, still lower voltages such as 24 volts, 15 volts DC, and even lower, are required for most control schemes. Most of the drive equipment commercially available in the United States requires one of these lower supply voltages. Some voltages, such as 575 volts, are used in various parts of the United States and Canada but are not as common. Interestingly, odd voltages do exist within certain facilities and industries, while typical frequencies are normally 60 hertz in North America and 50-hertz power outside North America.

In VFD systems, speed, torque, and horsepower directly relate to voltage, current, and watts with the basis for much of this found in Ohm's Law:

$$E = IR, \text{ or } Volts = amperes \times ohms$$

$$I = \frac{E}{R} \text{ or } Amperes = \frac{volts}{ohms}$$

$$R = \frac{E}{I} \text{ or } Ohms = \frac{volts}{amperes}$$

With so many VFDs being controlled by electric means, it is very necessary to have a basic understanding of many concepts and components of electricity as they interrelate to electronic drive and electric motor operation.

Voltage, usually shown as *V* or *E* is the *electromotive force* (EMF) which causes electrons to flow. Amperage—current—is expressed as *I* or *A* and is the actual flow of electrons in which the units of measure are amperes. Resistance *R*—ohms—is the opposition to current flow. Ohm's Law is adequate for DC circuit analysis and for some AC circuit analysis. Three-phase power, however, tends to be a bit more complicated.

The electric motor or prime mover in a variable-speed system is nothing more than an inductive load with some resistance in the overall circuit. The electrician can deduce, after measuring current, resistance, or voltage, where there might be a disparity and further troubleshoot to correct it. With variable-speed drives, power is defined by the type of drive in use, AC or DC. Electric power (*P*), the rate of doing work, is measured in watts and is expressed in formula form for DC circuits as $P = E \times I$. Alternating current circuits are analyzed differently when three phases of power are involved. Alternating current power has to be averaged and root-mean-square (rms) values provided to get proper results.

When discussing drives, power is discussed in terms of horsepower. However, the point is made throughout this book that VFDs should be sized by their load requirements and thus current. Horsepower and electric current are related, and, since *horsepower* is the prevailing term used in motor discussions, horsepower basics should be understood. Horsepower for a given DC circuit is equal to:

$$\frac{Volts \times amperes \times efficiency}{746}$$

Facilities that employ VFD and motor equipment today are trying to get the absolute best output from that equipment for the electricity that they purchase. Older mills and factories had enormous electric bills, and some still do. Now the mode is to have equipment as near to 100 percent efficient as possible. This is factored into automating any plant, new or old. The equation

$$Efficiency = \frac{746 \times output\ horsepower}{input\ watts}$$

can be used for AC circuits and is a good indicator of where a particular process or piece of equipment is, relative to its cost and productive output.

Frequency and Amplitude

Frequency is the quantity of electric pulses that are transmitted over a given period of time and is expressed in hertz (Hz). For example, most power in the United States is in the 60-Hz range (in Europe, it is 50 Hz). This means that every second there are 60 pulses of electricity through a given point in a wire. The typical waveform consists of the *frequency* portion (time based) of the wave and the *amplitude* portion (the magnitude). This wave is actually in sine-wave form, commonly referred to as the fundamental. For every electric wave, there must be a corresponding amplitude in order to provide any usable power. This amplitude is often in the form of voltage or current and is stated as the number of volts or amps at a given point in time. This wave has two time-based components called *periods* or *cycles*. A half period is one of the halves of the wave and is sometimes called a half cycle. This is illustrated in Figure 2–7 and is typical for AC.

In AC systems, the effective value of the wave is not the same value as would be found in DC circuits. The effective value is sometimes called the rms value. As AC power is generated by rotating turbines, there is a different value or strength of the voltage at every part of the 360-degree rotational cycle of the turbine. That explains the shape of the AC wave

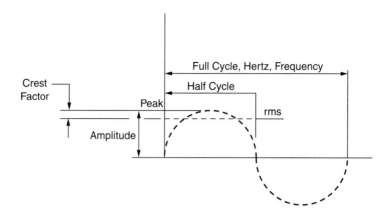

Figure 2–7 Sinusoidal waveform with frequency and amplitude components.

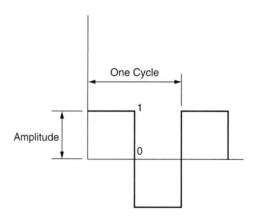

Figure 2–8 A typical square wave common to "on" and "off" states.

and also provides for the peak values of the wave. To find true, effective power, the rms value must be found to know the average work done through a given cycle. To find the effective value of an alternation or cycle, the peak value is multiplied by 0.707 (one divided by the square root of two). Also, taking rms voltage times 1.414 (the square root of 2) gives the corresponding peak value. A basic understanding of this concept is necessary for later recognition of what is going on with the output of a pulse-width-modulated (PWM), solid-state VFD.

Figure 2–8 illustrates a square wave, which is common to the output of many solid-state VFDs. Each pulse is a discrete on and off condition of a semiconductor or transistor. Its shape takes on the square-wave look due to this switching function (which explains the sharp rise to peak) rather than a rotational, changing state of amplitude. This frequency-and-amplitude pattern is a critical concept of the VFD. Sometimes called the *carrier frequency* of a PWM drive, this pulse pattern is further explored in a Chapter 3.

TABLE 2–1 **Various Frequency Bands**

Frequency Acronym	Range	Use
VLF Very Low Frequency	2–30 kHz	Timing Signals Industrial Controllers
LF Low Frequency	30–300 kHz	Navigational
MF Medium Frequency	300–3000 kHz	Land, Sea Mobile
HF High Frequency	3–30 MHz	Aircraft Mobile
VHF Very High Frequency	30–300 MHz	Radio and TV Broadcasting
UHF Ultra High Frequency	300–3000 MHz	Space and Satellite Communication
SHF Super High Frequency	3–30 GHz	Space and Satellite Communication

With frequency being the key component in a VFD's mode of operation, special attention must be given to how the PWM drive's frequency component fits in to the entire frequency spectrum. The radio-frequency wave that "carries" information is called the *carrier wave* or, in VFD realms, the *carrier frequency*. As can be seen in Table 2–1, the VFD's carrier frequency is actually low compared to the many ranges in use today. Some common values follow:

kHz = kilohertz (one thousand times per second)
MHz = megahertz (one million times per second)
GHz = Gigahertz (one billion times per second)

The audible frequencies are typically 30 Hz to 18 kHz for younger people with good hearing and 100 Hz to 10 kHz for older people. The high-frequency (HF) range is also referred to as the shortwaves.

Power Factor

An analogy for understanding power factor (PF) is graphically shown in Figure 2–9. A horse must pull a railroad car down the tracks but is unable to actually walk on the tracks. The horse must pull the railroad car from the side. The horse is pulling the car at an angle to the direction of the railroad car's travel. The power required to move the car on the track is the *real* power. The power put forth by the horse is the total *apparent* power. Unfortunately, not all of the horse's effort is used to actually move the car because the car cannot move sideways. This sideways pull of the horse is wasted, nonworking power and is called *reactive* power. The angle of the horse's pull is the ratio of *real* power to *apparent* power. Thus,

$$Power\ factor = \frac{real\ power}{apparent\ power}$$

As power factor approaches one, the reactive power approaches zero. If the horse is led closer to the center of the track, the side-pull angle decreases and the real power approaches

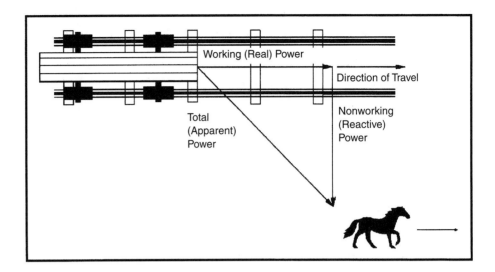

Figure 2–9   An example describing power factor (Reprinted from the U.S. Department of Energy's Office of Industrial Technologies BestPractices reference materials. Call the Information Clearinghouse at 1-800-862-2086 or visit the website: http://www.oit.doe.gov for additional information).

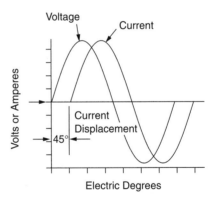

Figure 2–10   Current leading the voltage—leading power factor.

the value of the apparent power—an improving situation. This same power relationship exists with electrical requirements in facilities every day.

Electrically, the phase relationship between current and voltage is called *power factor*. As can be seen in Figure 2–10, the relationship between the current waveform and the voltage waveforms can be expressed in electric degrees. Thus, power factor, PF is the cosine of this displacement. The terms *leading* and *lagging* are often used in conjunction with power factor. They are typically defined by the *current* through the load. If the current leads the voltage across the load, then the load is said to have a *leading power factor*. Likewise, if the current lags the voltage across the load, then the load is said to have a *lagging power factor*.

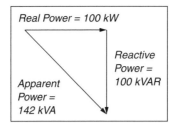

Real Power = 100 kW
and
Apparent Power = 142 kVA
then
Power Factor = $\frac{100}{142}$ = 0.70 or 70%

Figure 2–11 The power-factor triangle (Reprinted from the U.S. Department of Energy's Office of Industrial Technologies BestPractices reference materials. Call the Information Clearinghouse at 1-800-862-2086 or visit the website: http://www.oit.doe.gov for additional information).

The more reactive the load, the lower the power factor and, thus, the smaller the average power delivered. The more resistive the load, the higher the power factor and the greater the real power delivered. Low power factors are typically avoided, because high current is required to deliver any appreciable power. Higher current demand means higher heating losses, and, thus, the efficiency of the system suffers. This also means that the utility must have the basic equipment in place to deliver the current needed, even though the demand usage is not so. Therefore, the utility's costs are increased and that cost eventually filters back to the end user!

True power factor is made up of a displacement factor and a harmonic factor. Voltage and current waveforms *in phase* have respective zero crossing points that are the same. This condition allows for all the available power to be used as productive power. However, when the current gets *out of phase* from the voltage, the power is, for this half cycle, not as productive as it could be. This is the *displacement power factor*. The *harmonic power factor* is of a lessor degree, as it basically is the effect of the wave distortion in the same phase sequencing as previously described. Power factor is expressed in volt-amperes reactive (VARs) or kilovolt-amperes reactive (kVARs). This is the reactive power, a product of current, root-mean-square (a peak average), and voltage, root-mean-square. Power factor is proportionately equal to the watts divided by the volt-amperes. A diagram commonly used to illustrate power factor is the power-factor triangle shown in Figure 2–11.

Power-factor penalties imposed by the utility upon a facility vary. Whenever power factor, the ratio of working power to apparent power, is too low, the penalty charge is applied. The penalty charge is calculated and billed in these ways:

1. Kilovolt-ampere billing: Every kilovolt of apparent power or peak kVA supplied is billed, including reactive current. Thus, a demand charge of 2 to 3 dollar per kVA can be applied and then an additional primary kVA charge, a lump sum that includes a charge for every additional primary kVA.

2. Direct reactive energy charges: Reactive power is actually measured and a reactive energy charge is billed (so many cents per kilovolt-ampere-reactive-hour or kVARh). This charge can be 5 to 7 cents per every kVARh.

3. Demand billing with a power-factor adjustment: Bills are adjusted from the normal demand charge to account for the low power factor.

$$kW\ Billed = \frac{kW\ demand \times 0.95}{the\ facility's\ actual\ power\ factor}$$

In these cases, the power-factor penalty is imposed for power factor below 95 percent.

4. Excess kVAR demand charges: A maximum time period of reactive demand is allowed during a billing period, and any excess is subject to a surcharge.

It is obvious that the utilities are going to get their money. The problem is that those loads that need excitation and magnetism to function have a reactive-power issue about them. The reactive power does not perform "real work" and thus does not show up on the demand meter; however, the utility must have the transmission and distribution system large enough to provide it. This is why the utility feels justified in the extra charges. This forces facilities to put into place power-factor-correction equipment.

Capacitors

A *capacitor,* sometimes referred to as a *condenser,* is capable of storing electric energy. It consists of two conducting materials separated by a *dielectric,* or insulating material. The conducting materials get charged, one positively and one negatively, thus creating a potential between them. The size and type of material of the conductors, the distance between them, and the applied voltage determine the capacitance C in *farads* (one coulomb/volt). A *coulomb* is a basic unit of electric charge. One farad is a large amount of capacitance. Usually, common capacitances are in the microfarad range. Uses of capacitors include placing them into circuits for tuning purposes and even as ride-through circuits for stored electric power.

Capacitance can sometimes be good and necessary for a given circuit, and it can sometimes be a problem. The general effects of capacitance are that it:

1. Tends to keep voltage at a constant level

2. Tends to block DC current flow

3. Impedes low-frequency AC

4. Passes high frequency AC

Thus, whenever voltage must be sustained, a properly sized capacitor network can be included to accomplish that function. Likewise, since AC circuits have constantly changing values, the capacitor has a continuous effect; but it has little effect in a DC circuit. Also, capacitors are frequently used to impede low-frequency AC, to filter electric noise, and to reduce ripple currents. However, as can be seen in Figure 2–12, in which a given length of

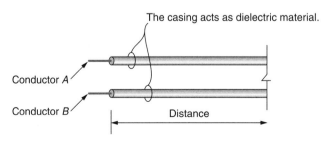

Figure 2–12 Power cables can exhibit some capacitive charging effects.

wire or cable actually takes on the appearance of a capacitor, the capacitive effect tends to produce a short circuit and reduces the output-signal strength. This can be very problematic when long cable runs are in place between electric components.

Inductors

Variable-frequency drives are basically made up of many electric components, eventually comprising a circuit. The principle of induction is in effect in that circuit. Depending on the drive construction, the induction component may be substantial. Whether the inductor is internal to a solid-state VFD or whether it is the electric motor prime mover, magnetic fields are present. *Inductance* is the ability of a conductor to produce induced voltage whenever the current varies. Inductors also oppose any change in current flow through a conductor. In analyzing inductance with respect to induced current:

1. A change in current flow will induce a magnetic field.
2. A changing magnetic field induces some voltage.
3. An induced voltage will oppose any change in current flow.
4. Inductors tend to pass low-frequency AC.
5. Inductors impede high-frequency AC.
6. Inductors allow DC to pass through.

Inductors are devices used to control current and the associated magnetic fields over a given period of time. Thus, the topic is pertinent to VFDs. Somewhere in the circuit will be an inductor. The sizing and the value of the inductor are also important. Inductance L is measured in units called *henries,* which equate to one volt/one ampere per second. Typically, an inductor consists of a coil of conducting material with a specific size and shape. This material is coiled around a core, most often of soft iron. For this reason, the inductor is sometimes called a *choke.* Most often, the inductor is called a *transformer* or *reactor.* It is often used to slow the rate of rising current and to suppress noise. Transformers are an integral part of the industry and drive electric circuits. Without these specialized inductors, there would be no available usable electricity in lower-voltage levels. A transformer's simplified construction is shown in Figure 2–13. Basically, transformers work on the principle of mutual inductance; hence, they are classified as inductors. For a transformer to work, there

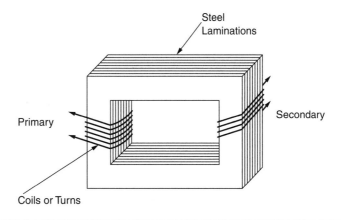

Figure 2–13   Simplified transformer (inductor) construction.

must be two coils, positioned such that the flux change that occurs in one of the coils induces a voltage across each coil. Typically, the coil that is connected to the electric source is the *primary* and the coil that is applied to the load is the *secondary.*

Transformers are rated in volt-amperes (VA) and kilovolt-amperes (kVA). The formula to calculate kilovolt-amperes is:

$$\frac{Volts \times amperes \times 1.73}{1000}$$

An example for a 60 horsepower application follows:

Given: 485-volt three-phase supply power, measured
60-HP VFD—ampere rating of 78 amperes

Find the kVA:

$$\frac{Volts \times amperes \times 1.73}{1000}$$

$$\frac{485 \times 78 \times 1.73}{1000} = \frac{65,446}{1000} = 65.45 \, kVA$$

A common rule-of-thumb practice is to multiply the horsepower rating by 1.1 to get an effective kVA:

$$60 \times 1.1 = 66 \, kVA$$

which covers the kilovolt-ampere requirements of the application. However, the heating that the transformer may undergo due to the waveform distortion and nonlinear load created by the VFD may cause some transformer derating. Thus, a slightly larger transformer may be required. Also, transformer manufacturers typically have standard sizes available from which to choose. It is usually costly to have a custom-size transformer built. The VFD used will influence the amount of derating or service factor the transformer manufacturer requires. This derating and K-factor information is disussed later in this chapter. Always consult with a transformer manufacturer when applying a drive-isolation transformer (DIT).

There are basically two types of transformers: dry-type and oil-filled transformers. There are also iron-core types, air-core types, and units that are called autotransformers. An *autotransformer* is a type of transformer that has one winding common to both the input and the output. Instead of having isolation between the coils and employing the traditional two-circuit principle, one winding is used and higher kilovolt-amperes can be transformed. The autotransformer is often less expensive than a full-isolation or voltage-matching transformer. This is the case when voltages that are to be matched are nominally not that far apart. Additionally, transformers can be used as isolation devices or as step-up or step-down (voltage-matching) devices in an electric circuit of a variable-speed-drive system. Often times, it is desirable to include an isolation transformer with similar primary and secondary voltages mainly to isolate one section of the electric system from another. This way there is no direct physical connection, and, thus, conditions such as ground faults can be protected against traveling throughout an electric system and destroying components. Also, some high-frequency AC is blocked in an inductor, thus utilizing the isolation transformer as somewhat of a noise suppressor. However, do not assume that an isolation transformer takes all electric noise from a circuit. Also, the step-down or step-up transformers used to match voltages in a given system also offer isolation. Much of today's electronic equipment requires 460/480-volt, 230/240-volt, or 120/110-volt power to function properly—thus the need for transformers.

When cost is a factor, voltage matching is not an issue, and supply-line disturbances are minimal, a *line reactor* can be used in place of the isolation transformer. This inductor is dedicated to adding line impedance and is usually not as substantial as an equivalent transformer. It weighs less and is less expensive. The line reactor is also used on the output of an electronic drive to protect the output transistors in times of short circuit on the output line and to reduce the standing wave condition whenever long cable lengths are seen on the output to the motor. This condition is reviewed in Chapter 4 when installation of AC VFDs is discussed.

Most transformers used today have to be designed for both linear and nonlinear loads. Incandescent lighting and line-started motors are examples of linear loads. Nonlinear loads are typical of any electric equipment that has a power-switching component associated with it. In addition, nonlinear loads include electronic lighting ballasts (fluorescent lighting), arc-welding equipment, copiers, computers, printers, and so on. Thus, nonlinear loads have put an additional demand on transformers. This increased demand can be addressed in new installations, but adding rectifying and phase-controlled converters to existing transformers is a topic in need of investigation. Nonlinear loads demand current from the utility that creates higher frequencies. The waveform or shape of the load current is no longer defined by a single frequency. It is a complex shape that contains many frequencies.

High-frequency currents will attempt to flow through the surface area of a conductor. When the cross-sectional area of that conductor becomes too restrictive, the conductor eventually becomes hot. When the conductor is packaged inside layers of wire, as is the case within a transformer, the temperature of the transformer begins to rise and hot spots are created, which will ultimately lessen the life of the transformer. Transformer manufacturers, knowing when nonlinear loads are going to be present, take this into account when sizing the transformer. The rule of thumb is to increase the size of the transformer, sometimes by 10 percent, which in effect derates a transformer with a higher rating. Using a larger transformer does not always guarantee that it will run at a lower temperature. A larger transformer will use wire with a larger cross-sectional area. Increasing cross-sectional area does not provide a proportional increase in surface area. Harmonic currents can cause hot spots in oversized transformers. Harmonics associated with drives also increase the eddy current losses in a transformer; this loss is difficult to predict. Thus, there are two major issues when dealing with nonlinear loads such as those generated by electronic drives.

One issue deals with the hot spots within the transformer that is going to run nonlinear loads and involves shaping the wire so that there is as much surface area as possible to offer the least resistance for the higher-frequency currents. The other is to attempt to specify the amount of harmonics by use of the *K factor*. Today's transformers are now classified with a K factor. The K factor defines the transformer's ability to handle harmonic currents while operating within the thermal capability of the transformer. Linear-load transformers are classified with a K factor of 1. Transformers with a K factor of 4 are suitable for moderate levels of harmonic currents. A K factor of 13 is suitable for greater levels of harmonic currents, as might be seen with current source drives. Likewise, the typical PWM drives used today will carry a K-factor rating of 6. The K factor is merely an attempt by industry to define and address the issue so that equipment lasts longer and catastrophic failures are kept to a minimum.

Another issue when dealing with nonlinear loads is the rate of rise in the temperature of the transformer. There is a constant that is attributed to transformers and other peripheral equipment that is subjected to higher electric currents. The rate of temperature rise is defined as the allowable rise in temperature over the ambient temperature when the device is fully loaded. *Ambient temperature* is that of the surrounding air at the device itself, which acts as the heat sink for cooling. When comparing two or more different transformers for the same application, a rule of thumb is to compare the weights of the different units. If one has less iron and copper (or aluminum) winding content, then it will show in the actual weight. This may come into play when a transformer has to be provided in an application where harmonics may be present. These harmonics will create eddy current losses within the transformer. The severity is hard to predict, but a transformer could overheat. To deal with this phenomenon, the K-factor value is now attributed to transformers and the amount of harmonics that they can handle. The K factor is sometimes inappropriately called the *form factor*. The term *form factor* refers to the ratio of root-mean-square current to average current. The form factor, more often, is the effect of rectifiers on motors. However, anytime there are devices performing power conversions (AC–DC, DC–AC, etc.), there will be some sort of distortion to the waveform somewhere. That is why transformers have to be analyzed with regard to their K factor.

Semi-conductors

Today's VFD of choice is a solid-state device. For reasons revealed in Chapter 1, more and more of this type of equipment is becoming commonplace. The piece(s) of hardware physically installed within a solid-state VFD that switches, turns on and off, conducts and does not conduct, is the *power device.* There are many types of power devices incorporated into electronic-drive-rectifier and inverter circuits. They can be built as modules for ease of installation, or they can be left as individual units. New power devices are being developed, and new ways of using them are also being explored. These devices become the main component within any solid-state AC or DC drive. Not being able to change electric energy for specific motor-control purposes makes the electronic drive useless. Power semiconductors and similar devices have been used in electronic-drive equipment for years. Often called thyristors, these devices can be silicon-controlled rectifiers (SCRs), gate-turn-off devices (GTOs), and the device that preceded them—the vacuum tube.

The *vacuum tube* or electron tube, dating all the way back to the late 1800s, provided the standard method for rectifying AC. It can also be referred to as a *diode,* as it acts as a one-way valve to control the flow of electrons. Its basic construction is that of a glass or metal enclosure from which the air has been evacuated—thus, the name vacuum tube. There are pins at the base of the unit where connections must be made to the *cathode,* which is typically made from tungsten and supplies electrons via a filament that heats the cathode, allowing electrons to be emitted. The *anode,* called the *positive plate,* collects the electrons. Additional electrodes called *grids* control the overall emission.

Another power device used to convert AC to DC is an electron tube called the *thyratron,* which is a tube that is filled with gas and utilizes three elements. However, with semiconductor technology and advanced solid-state electronics, electron tubes have been all but forgotten. There are old installations still dependent on their rectification, but the reality is that there are no spare parts readily available anymore and no one wants to support this old technology. Industrially, the *ignitron* was a device used to rectify currents at higher-current ratings, but, alas, tube technology for practical purposes is a thing of the past.

Vacuum tubes evolved into thyristors for power rectification. *Thyristors* are types of transistors in which there are several semiconducting layers with corresponding positive-negative (*p-n*) junctions. The thyristor is a solid-state version of the aforementioned thyratron device. The most common thyristor is the silicon-controlled rectifier (a four-layer device, p-n-p-n). The SCR is still a fairly common device used in power rectification. It is similar in operation to that of a diode (a two-layer device), but its primary difference is that it is controllable. The SCR can block the flow of current in the reverse direction, much like the diode, but it can also block the flow of current in the forward direction. The SCR has a blocking state and a conducting state. In its blocking state, no current is allowed to flow through it. Likewise, in its conducting state, it acts like a switch that is closed. The SCR receives a small current signal— the *gate signal*— to be triggered into conduction. The SCR will keep conducting until the gate—the all-right-to-conduct—signal is removed, and the current flow reduces to zero. This turn-on and turn-off function of the SCR allows for extremely good control and very small losses in terms of current leakage. The SCR, while conducting, has a very good forward voltage-drop value, which means that large amounts of current can flow through it with very little in the way of energy loss.

There are many different SCR devices. There are complete modules, stud types, and hockey puck versions. The hockey puck design of an SCR is most common; it looks like a hockey puck but is usually white in color and has two leads for receiving its gate instructions. Very high values of current can be run through the SCR, thus making it the rectifier of choice in higher-horsepower applications. There are additional ratings attributed to SCR design. The peak inverse voltage (PIV) is typically in the 1400-volt range. The PIV rating is sometimes called the peak reverse voltage (PRV), which means that a device will only block a certain amount of voltage in the opposite direction. If this voltage rating is exceeded, then the device can be destroyed. Another factor when selecting devices is the proper heat dissipation. The surface area for heat dissipation is important.

GTO thyristors are power semiconductors with a self-commutating ability. This means that, once commanded, the GTO will turn on and turn off repeatedly without having to have the gate signal removed or applied. These devices are available in high current ratings and have high overcurrent capabilities, thus making them very suitable for larger-horsepower applications. Their turn-on and turn-off times are good, and the speed at which they switch is adequate for the typical higher-horsepower applications in which they are used. However, a major drawback to using them is that their cost, when compared to similar phase-controlled devices, is five to six times higher than that of a conventional SCR device.

Modern solid-state VFDs usually incorporate the *transistor*. These solid-state components have been around in some form since the early 1950s but really have gotten popular in the past decade. Basically, they are made up of different semiconductor materials, sometimes with arsenic or boron in conjunction with silicon. The way the electricity moves through the silicon is directly related to the amount of arsenic or boron contained. What transistors provide is fast switching capability for a relatively low cost. The general types of transistors are: the bipolar transistor, the field-effect transistor (FET), the insulated-gate-field-effect transistor (IGFET), and the insulated-gate-bipolar transistor (IGBT).

Bipolar transistors can be found in oscillators, in high-speed integrated circuits, and in many other switching circuits such as those in VFDs. Bipolar transistors are available in rated currents much lower than thyristor-type devices, six to seven times lower. However, these transistors can be paralleled in operation to achieve greater current-carrying capacity, with the drawback being higher cost for this design. Self-commutating devices, they cannot withstand too much in the way of overcurrent conditions. Bipolar-transistor switching speeds, for their time, were in the 2 kilohertz to 4 kilohertz range, thus making it a very good switching device for its relative cost, even though it had somewhat marginal turn-on and turn-off capabilities. However, with the introduction of faster switching devices, bipolar transistors are starting to be replaced. One such fast-switching device used in place of bipolar transistors is the metal-oxide semiconductor–field-effect transistor (MOSFET). It is a self-commutating device, which is not too costly to incorporate into a system. The present drawback is that MOSFETs are not available in current ratings much above 20 amperes. This limits its applications, even though it has switching capabilities in the 100-kilohertz range, good overcurrent capability, and very good turn-on and turn-off conditions.

The IGBT, is very common in VFDs today. This self-commutating device is available in 300-ampere current ratings, has good turn-on and turn-off ability (needs only 3 to 5 volts of energy to turn on), and has switching speeds of up to 18 kilohertz. It is cost effective to manufacture and can be implemented into an electric circuit at relatively low costs. With costs and performance driving the semiconductor industry, more efficient versions of IGBTs are coming out every year, which is why VFD designs change so often!

Most solid-state VFD designs now incorporate a diode bridge as the rectifier. A diode is a solid-state rectifier, which has an anode—the positive electrode—and a cathode—the negative electrode. These nodes allow electricity to flow in one direction only. Diodes are commonly used to convert AC voltage to DC. The diode, in effect, acts as a blocking valve for electricity. This diode might be used for voltage regulation. A reverse-operating version of the conventional diode is the zener diode. This is sometimes called a breakdown diode, which can allow reverse currents under breakdown conditions. Sometimes this is desirable in regulating voltage.

CHAPTER 3

AC-Induction-Motor Theory and Variable-Frequency-Drive Basics

AC-Induction-Motor Theory

To understand how and why a variable-frequency drive (VFD) functions the way it does, it is important to have a complete understanding of the object it will control—the *AC induction motor*. Knowing the AC motor basics will allow for better setup and trouble-shooting of the VFD. After all, the motor and drive are part of the same electric circuit, as is seen in Figure 3–1. Once in that circuit, they are one—dependent on each other to successfully operate and move the load at a desired speed. Also, as the VFD is an electronic assembly, knowledge of how a motor uses voltage and current to produce speed and torque will prove valuable to understanding the overall control circuit.

AC-Motor Construction

AC induction and repulsion motors are similar in operation and construction. Repulsion motors, induction motors, and repulsion-induction motors all have a stationary element—the *stator*—and a rotating element—the *rotor*. Basically, the *induction motor* works on the principle of changing the state of electromagnetism around or within a magnetic field to achieve motion. In this way, it acts as a rotating transformer. The *repulsion motor,* which is close in operation, contains two magnetic fields of like polarity; thus, they oppose each other and cause motion. A *repulsion-induction motor* is the hybrid and actually contains brushes, a commutator, and a wound rotor. It uses the brushes and commutator to get hard-to-start loads going. Once it reaches 65 percent to 75 percent speed, the brushes are lifted off of the commutator by centrifugal force, and it continues running as a squirrel-cage-type motor.

A vast majority of industrial and commercial-duty *induction motors* are three-phase supplied and are sometimes called polyphase motors. The electric action of this type of motor, particularly the squirrel-cage type, is likened to a transformer with a shorted secondary (Figure 3–2). The motor is much like a step-down transformer in that the primary is equivalent to the windings of the motor. This is where the three-phase AC power is fed. The secondary is likened to the rotor, or armature. Fewer secondary turns and a heavier gauge of wire equate to higher value of current induced by the primary with the maximum current created when the secondary is shorted. The squirrel-cage induction motor's construction is shown in Figure 3–3. Its construction resembles that of a squirrel cage—thus, the name. The rotor is built up from steel laminations. Each lamination has a set of slots around its

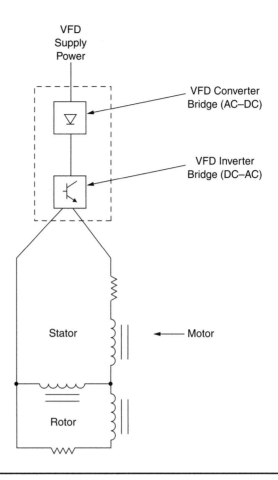

Figure 3–1 The motor and the VFD as the same circuit.

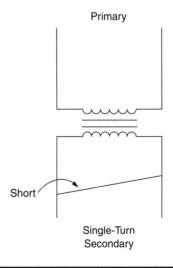

Figure 3–2 An AC induction motor acts as a rotary transformer with a shorted secondary.

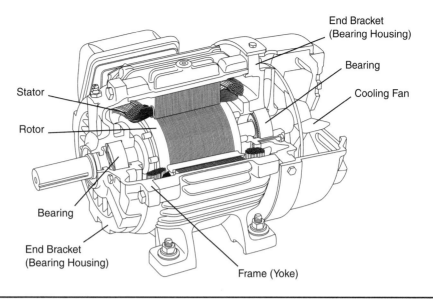

Figure 3–3 An AC induction motor's construction (Courtesy of Siemens Energy and Automation, Inc.).

perimeter. Into these laminated slots go the long rotor bars. These rotor bars are usually made from copper or aluminum and have various cross sections to increase impedance by creating eddy currents. Traditionally, a squirrel-cage-type motor has had a disadvantage because of its fixed-rotor design. This disadvantage is in actual starting torque. Different rotor-bar designs can provide the extra impedance to better control the rotor flux at low speeds. At high speeds, this is not an issue.

Induction Motors and Control

AC motor speed change can be accomplished three ways: (1) Change the number of poles in the motor; this means separate windings; (2) Change the slip characteristics of the motor; this is done with varying resistors such as is done with a wound-rotor motor or by varying the stator voltage; or (3) Change the frequency of the power supplied to the motor. The third option over time has become the method of choice—hence, the variable-*frequency* drive.

Synchronous speed is the speed of the rotating electric field within the induction motor. It is not the actual motor rotor speed. To calculate synchronous speed, the following formula is employed:

$$Synchronous\ speed = \frac{120 \times frequency}{number\ of\ motor\ poles}$$

For example, a motor with 4 poles and supplied 480-volt AC power at 60 hertz has a synchronous speed of 1800 rpm. Table 3–1 shows the corresponding speeds for motors with different pole counts. However, it is very important to note that once a motor is loaded, it cannot reach that synchronous-speed value in rpm. The difference between the

TABLE 3–1   **Induction-Motor Speed/Poles**

Number of Poles	Synchronous Speed in rpm (50 Hz)	Synchronous Speed in rpm (60 Hz)
2	3000	3600
4	1500	1800
6	1000	1200
8	750	900
10	600	720

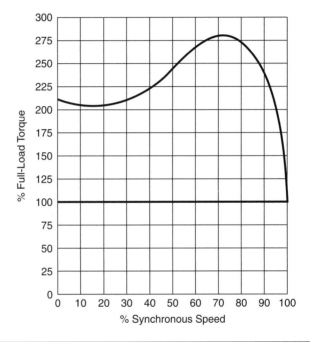

Figure 3–4   Typical speed versus torque curve for a NEMA design A motor.

synchronous speed and the full-load motor speed is called *slip*. An example 4-pole motor may have a synchronous speed of 1800 rpm and roughly 3 percent slip. Its full-load speed (as nameplated) is then 1800 minus 55, or 1745 rpm. Induction motors are often classified by their slip characteristics. Further evaluation of this characteristic can be found in speed-versus-torque performance. An AC motor, when applied a load, has a tendency to slip. This phenomenon allows for the motor to request more current to maintain driving the given load at the desired speed. Some motors have more slip than others. Typical values of slip are 2 percent to 3 percent of synchronous speed; this is common with NEMA (National Electrical Manufacturers Association) A (Figure 3–4), NEMA B (Figure 3–5), and NEMA C (Figure 3–6) design motors. Some values can get as high as 5 percent to 6 percent for special high-starting torque applications. These higher slip values would be found in NEMA D (Figure 3–7) and other design motors.

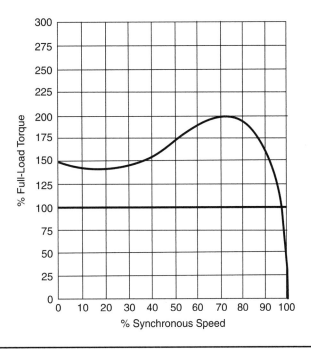

Figure 3–5 Typical speed versus torque curve for a NEMA design B motor (Courtesy of Siemens Energy and Automation, Inc.).

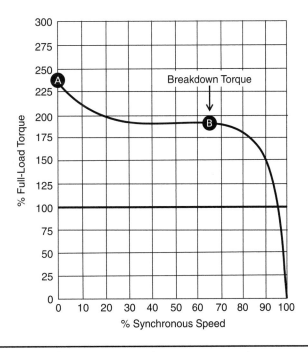

Figure 3–6 NEMA design C motor with 225 percent starting torque at point *A* (Courtesy of Siemens Energy and Automation, Inc.).

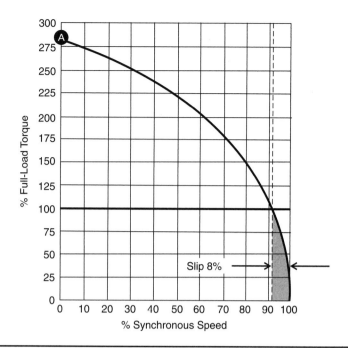

Figure 3–7 Typical speed-versus-torque curve for a NEMA design D motor (Courtesy of Siemens Energy and Automation, Inc.).

Variable-frequency drives are often employed to control and operate a wide variety of induction-type AC motors. Many times the VFD *can* operate a motor other than the traditional squirrel-cage type, but there may be limitations such as not enough starting torque, a motor running hot, or even not being able to achieve full speed. Consult with both the drive supplier and the motor supplier whenever these conditions present themselves. Other types of induction motors include the capacitor motor, the shaded-pole motor, the wound-rotor (or slip-ring-rotor) motor, and other polyphase motors.

The capacitor and shaded-pole type are single-phase units, limited in size. The *capacitor motor* works on the principle of storing energy for starting and can be obtained in horsepower sizes up to 20 horsepower. The *shaded-pole motor* utilizes salient poles, which are copper loops acting as the starting winding. They are also called shaded poles—thus, the motor's name. The shaded-pole motors are very inexpensive and do not get much above $\frac{1}{3}$ horsepower.

Wound-rotor motors are another type of induction motor that can be found in horsepower sizes from fractional to several thousand horsepower. The equivalent electric circuit can be found in Figure 3–8. The wound-rotor motor contains many groups of copper coils. The stator windings equal the number of poles so that a rotating magnetic field is produced. A bank of resistors is also part of the wound-rotor-motor circuit. The resistors are used to limit the current to the rotor, thereby causing slip, which provides the torque necessary to drive a given load. The parts of a wound-rotor motor are shown in Figure 3–9. Typically, the wound-rotor motor has been used more for better torque regulation than for

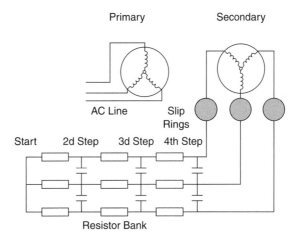

Figure 3–8 A wound-rotor motor's electric circuit.

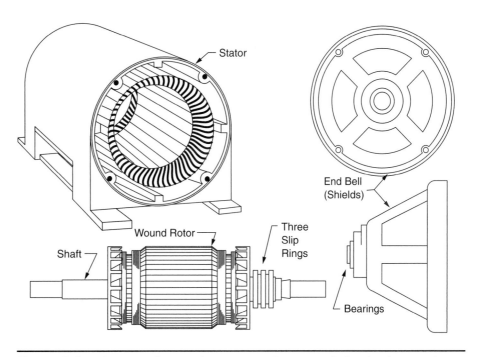

Figure 3–9 The parts of a wound-rotor motor (Courtesy of Alerich, *Electricity 4,* Delmar Publishers).

speed regulation. It has been used for years in the crane-and-hoist industry because it can provide high starting torque. The slip rings are often tied together in the wound-rotor motor, which allows it to operate as an induction motor and be controlled by an electronic VFD. Sometimes extra resistance is required in the rotor circuit to get it to start as an induction motor.

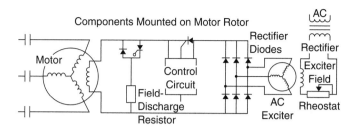

Figure 3–10   Simplified circuit for a brushless, synchronous motor (Courtesy of Alerich, *Electricity 4,* Delmar Publishers).

Synchronous Motors and Control

Synchronous motors are another class of AC motors. Variable-frequency drives of the PWM, transistorized design are not well suited to control a synchronous motor. A simplified circuit for a brushless, synchronous motor is shown in Figure 3–10. Basically, the induction and synchronous motors differ mainly in the design of their rotors. The stator function is virtually the same in each type motor. With synchronous motors, the rotor and the magnetic field are running at the same speed; with induction motors, the magnetic field and the rotor are running at different speeds. All of these "non-squirrel-cage-type" motors are used on various applications in industry and are often requested to be fit with a VFD. Thus, knowing how they function is important to the applicability of a drive.

Because VFDs are not the controller package to use with synchronous motors, there are specialized electronic controllers that can start and run a synchronous motor. Synchronous motors run at a constant speed regardless of the load changes. They run from a fixed-frequency AC source and run at the same speed as the source frequency and as dictated by the number of poles in the synchronous machine. The synchronous motor's rotor has poles that "synchronize" with the rotating magnetic field. Many synchronous machines have low synchronous speeds and thus have 10, 12, and even higher counts of poles. These motors are used quite often in compressor and pump applications. They can be provided in very high horsepower configurations and are well suited for steel and aluminum mill environments. There is often the need to start the synchronous motor with a DC generator so the poles can eventually lock in with the field, especially whenever a variable-speed drive is used. The load on the synchronous motor must be within its rating—within the electromagnetic force generated between the rotor and the magnetic field. Higher loads will cause the motor to stall.

Nameplate Data

No matter which type of motor is selected—induction or synchronous—there will be valuable information supplied on its *nameplate,* which hopefully is intact, not painted over, and legible. From this basic data, one can determine many characteristics about the motor's expected performance and will be able to size other electric components on the circuit. A typical induction motor nameplate is shown in Figure 3–11; various examples of necessary data derived from that nameplate are shown in Figure 3–12. Inverter-duty or inverter-fed motors are discussed at length later in this chapter, and their nameplating is especially important.

Serial #	120356RC-69	**HP** 30		**Frame**	286T	
Phases	3 **Hz** 60	**Voltage**	460/230	**rpm**	1755	
FLA 36.2/72.4	**SF** 1.15	**NEMA Design**		B	**Encl** TEFC	
Code G		**Insulation**	Class H	**Ambient Temp**	40°C	
Duty	Inverter Duty—Suitable for 6:1 Turndown **Misc** C-Face					

Figure 3–11 Typical nameplate data from an AC induction motor.

From nameplate data, the following can be found:

1. The subject motor is an inverter-duty motor, which can be applied to the output of a VFD. The safe-speed range is 6:1, or from 60 Hz to 10 Hz operation while running on the VFD. The motor's cooling is adequate for this turndown.
2. The insulation class is H and is suited for a temperature rise of 180°C.
3. HP is 30, and speed (rpm) is 1755. Torque can be found,

$$T = \frac{30 \times 5250}{1755} = 89.75 \text{ Foot Pounds}$$

4. The NEMA design of this monitor is B, which indicates the starting and inrush current values. Along with the Code G, branch-circuit protection and fusing can be selected.
5. The 286 frame size indicates a foot-to-shaft center of $\frac{28}{4}$, or 7 inches.

Figure 3–12 Useful information derived from an AC induction motor's nameplate.

NEMA Ratings

The National Electrical Manufacturers Association (NEMA) has provided for some standardization in the motor industry. This standardization includes physical data, mechanical sizing information, and electric characteristics. For instance, three-phase, AC induction motors have specific ratings for speed and torque. Each of the NEMA designs has a different characteristic for starting current, locked-rotor current, breakaway torque, and slip. The most common designs are NEMA A, B, C, and D. Each has a distinct speed-versus-torque relationship and different values of slip and starting current.

The most common design is the NEMA B motor. Its speed/torque curve was shown in Figure 3–5. Its percentage of slip ranges from 2 percent to 4 percent. It has medium values for starting and locked-rotor current and a high value of breakdown torque. This type of motor is very common in fans, pumps, lighter-duty compressors, various conveyors, and

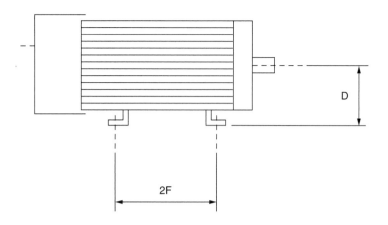

Figure 3–13 Critical dimensions derived from the AC motor's frame number.

some lighter-duty machines. It is an excellent choice for variable-torque applications. The NEMA A motor has similar characteristics to the NEMA B motor. It typically has a higher value of locked-rotor current, and its slip can be higher. NEMA C motors are well suited to starting high-inertia loads. This is because they have high, locked-rotor torque capability. Their slip is around 5 percent, and their starting current requirement is average. The NEMA D motor is that motor found in heavy-duty, high-inertia applications. It has high values of slip, very high locked-rotor torque capability, and usually a base speed of below 1800 rpm. Typical applications include punch presses, shearing machinery, cranes, and hoists. Whenever applying a VFD to NEMA C or NEMA D motors, attention must be given to the slip and full-load-current ratings of these motors to get a proper size match (drive current rating and motor current rating).

A motor is often supplied based on its frame size. Motor manufacturers have their own designations when the horsepowers exceed a certain level. NEMA has provided for some standardization in motor sizes up to the 500-frame series for AC motors. Above this rating, manufacturers of motors can classify their motor frames with their own designations (these are referred to as "above-NEMA" ratings). This means that for a given horsepower rating of a motor and for a given base speed there are standard frames that most manufacturers adhere to. For instance, there is a certain foot-to-centerline-of-shaft dimension and a certain frame diameter (see Figure 3–13). Dividing the first two digits by 4 usually indicates the foot-to-centerline-of-shaft dimension (324 frames have an 8-inch dimension, $\frac{32}{4} = 8\,inches$). This helps the machine designer in locating the motor physically to the machine and in comparing a motor sized by one manufacturer with one by another. There are other issues to consider based on the duty cycle, service factor, ambient conditions, and torque requirements. Many times, different frame sizes will be offered by different manufacturers for the same application and horsepower. Fully analyze what all the impacting factors mean and which are most important.

Motor Cooling

Applying VFD controllers to motors means that the speed of that motor is going to be changed, usually lowered from full speed. This ultimately means that the fan attached to the motor shaft will turn slower and cooling air will *not* be delivered. Depending on the motor loading, special attention must be given to how the motor is to be cooled. Addressing cooling needs of the motor starts with the application type. Whether it has constant or variable torque, how low in speed the motor is to run, and what type of motor exists or will be selected are all very important. Whenever electric current is passed through the electric motor, there is a buildup of heat. The laws of physics predict that there will be heat produced. The amount of heat produced is a function of the work, or loading, done by the motor; the type of waveform of the actual electric signal to the motor; and the eventual changes due to bearing wear and friction. Premium or high-efficiency motors provide even better use of the electric energy to get more work than losses out of the circuit. Special attention to rotor-bar designs and cross sections, better conducting materials, larger-size wire, thinner laminations, and tolerances in the motor's air gaps can result in better efficiency (handling the heat buildup) out of the electric motor.

A motor that runs fully loaded or sometimes overloaded will invariably require a greater amount of current. This could affect its heat content, especially with respect to the duty cycle and the speed range. Another factor that affects motor heating is the incoming power signal itself. A pure sine wave will provide a known value for calculating motor losses. However, when that signal is subject to spikes, noise, or other line disturbances, the motor suffers as losses go up and less of that incoming power gets used for flux and torque production. Additionally, over time, bearings wear as do other driven components in the drivetrain. This may cause the motor to perform extra work; thus, there can be extra heating. Getting the heat away from the motor is very important.

Many motors are sized for a particular application or horsepower rating so that the heat produced from the current can be accepted by the metal content of the motor. Normal convection and radiation dissipate the heat with the aid of an internal mixing fan. These motors are classified as open drip proof (ODP) or totally enclosed nonventilated (TENV). Other electric motors incorporate a fan blade that rotates at the same rpm as the motor shaft since they are physically attached to the shaft. This fan blows air across the outside of the motor, thus cooling it as it runs. However, if a VFD is used, it is apparent that the lower in speed the motor is made to run, the slower the cooling fan will run also. Thus, a dangerous condition can exist with respect to heat buildup in the motor. Fortunately, the drive will protect the motor by tripping on an overtemperature fault, which is mainly a nuisance. These motors are called totally enclosed fan cooled (TEFC).

Thus, should the motor heating be too great compared to the rate of heat dissipation as with ODP or TENV designs or if the motor rpm are not adequate to move the heat away in a TEFC design, then auxiliary cooling measures must be taken. If not, the motor will either damage itself or, if motor thermal protection is built in, stop running. Either condition is undesirable, and up-front care can be taken to eliminate having to go through these circumstances. Therefore, steps can be taken when applying AC or DC drives to motors to know ahead of time how low in speed the motor will actually run and under what type of load. One approach is to size the motor with a service factor. Another is to simply go up in horsepower, which is how motors often are sized. This may put a motor into a larger frame designation, thus probably making it weigh more and allowing it to handle a greater

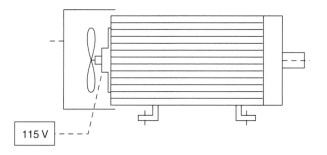

115 V

Figure 3–14 Separately powered auxilliary blower for added cooling at low speeds.

amount of heat. A service factor of 1.15 means that the motor has 15 percent more capacity when operating conditions are normal for voltage, frequency, and ambient temperature. This 15 percent extra capacity means that the motor is built and sized to handle operation from an electronic drive, the duty cycle is severe, or the loading-and-speed range is moderate. Two scenarios often emerge: (1) One that occurs often is that a motor is selected for a particular load and duty cycle. Someone decides to play it safe and asks for a service factor of 15 percent. Then another player bumps the horsepower rating up by a factor of two. Pretty soon the application has a motor well oversized for the loading, and, therefore, energy is being wasted with the initial cost of the motor high. (2) Another is the exact opposite. A motor is selected for a variable-torque load with a 4 to 1 speed range. The next thing that happens is that the motor is running full current down to 10 percent speed, and a heat problem may exist.

Fortunately, the other option for motor cooling is one that can be added on later (in those conditions where the motor was applied for the loads and speeds not originally specified) when a heating problem emerges. This is the auxiliary blower; it is a separate fan motor that is mounted on the main motor as shown in Figure 3–14. This auxiliary fan motor is much smaller and runs full speed all the time simply providing moving air across the larger motor to take the heat away and cool it. All that is required here is a starter, installed usually at the drive, and a smaller, full-speed motor with the fan.

Another method of cooling a motor is to duct cooled air to the motor. This is appropriate in an environment that is harsh or in which explosive gases may be present and a separate auxiliary fan motor is not attractive nor practical. This way, clean, cooled air can be provided. Other means of motor cooling are possible but are usually more expensive. Water or liquid cooling are some other methods used. Often times, the environment where the motor is to reside dictates the type of cooling method chosen. Whenever a VFD is controlling a motor, special attention must be given to cooling that motor.

Motor Protection

In addition to cooling the motor, attention must be given to protecting the motor. The more expensive the motor, the more protection should be considered. After all, it is an investment and should last several years when properly maintained. Motors should be equipped with thermostats that cause the supply power to cease once a predetermined motor temperature is reached. A motor is often running off of an electronic drive that can monitor the current being sent to the motor. The drive can shut off power to the motor if

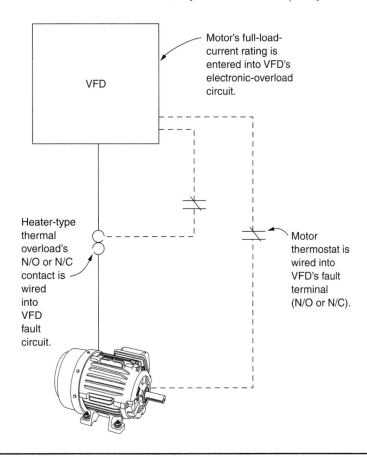

Figure 3–15   A well-protected motor.

it detects an overload condition exceeding a certain length of time. Figure 3–15 shows different motor-protection schemes. Variable-frequency drives (most often a drive but can be a starter/motor protector) have an electronic-overload circuit built in. It can monitor the current going out to the motor and shut off power to the motor in the event of excess current sensed over a fixed period of time. Another method of motor protection is by thermal-overload relays. Here, this overcurrent device is selected at a predetermined size and current value to cause disconnection of the motor from the power supply when that condition exists. This device does not protect itself, but, rather, it protects the load, which is the motor. Lastly, the motor thermostat, the simplest form of motor protection, can be employed. This contact, when opened, shuts power off to the motor. Sometimes these contacts are used in addition to other motor-protection means (i.e., going into a electronic drive's fault circuitry). Extra protection can never hurt.

Motor Windings and Bearings

The AC induction motor, as it is constructed, exemplifies a product whose design has evolved over the years based on the need for extra protection, to gain extended motor life, and to be very efficient. Special care is taken when the *windings,* which are basically turns of copper or aluminum wire, are incorporated into the motor package. Each turn or coil of wire must be placed precisely or else a premature failure in the motor will occur. Often times, a motor is required to be *form wound,* which means the even, extra special placement of the turns in the windings must occur. The term *random wound* is used for describing the more common method of winding. As the windings are being wound, insulating tape is applied to further isolate the windings from each other. Then the entire network of coils is coated, or sometimes dipped into a tank, with a varnish. This seals the insulation on the windings and adds to the life of the motor. Remember, any deviations or imperfect turns of wire can cause extra heating and, over time, will degrade the motor's torque-producing capabilities and eventually cause overheating.

As will be discussed in Chapter 4, another important piece of the motor package is the *bearing* arrangement. The front and the back of the rotor must rest on sets of bearings. Often times, premature wear at the bearings in a motor can cause a machine to shut down. Most of the time, the early failure of the bearings can be attributed to the actual mounting of the motor itself. If there is side loading or if the motor is coupled in a manner such that the shaft is somewhat off center, then unwanted forces begin to act on the bearings, wearing them out faster. Heavy-duty bearings should be requested when the motor mounting is overhung or side loaded. Additionally, good maintenance and lubrication of the bearings (if the size of the motor allows for lubrication—many motors have sealed bearings) will lengthen their life. Some motors can be supplied with oil-mist systems, which perform lubrication continuously. These systems are prevalent when motors must rotate at extremely high speeds.

Motor Insulation

Motor insulation is a very critical issue with AC induction motors fed by VFDs. Motor insulation is that nonconducting material that basically separates current-carrying components within the motor from each other. This is often in the form of insulating tape, which is applied prior to any dipping or coating of the windings. Motor insulation has been given ratings based on *temperature maximums,* allowable temperature rise within the motor. These values are shown in Table 3–2 and reflect choices made due to the ambient temperatures and the motor operation. Insulation ratings are a good gauge when applying motors to VFDs. Variable-frequency drives do promote extra heating in the motor due to a nonsinusoidal waveform. This extra heating, depending on its severity, can lessen the life of the motor, although this phenomenon may be somewhat overexaggerated. Many factors enter into the life expectancy of a motor—starting currents, quantity of starts in its life, shock loading, and so on. It is always better to get the better grade of insulation whenever possible and practical. This can only help to lengthen the motor's life, whether or not an electronic drive is controlling it.

TABLE 3–2   **Motor Insulation and Temperature Rise (°F, °C)**

Class A	239° F	115° C
Class B	266° F	130° C
Class E	248° F	120° C
Class F	311° F	155° C
Class H	356° F	180° C
Class N	392° F	200° C
Class R	428° F	220° C
Class S	464° F	240° C

Inverter-Duty Motors: NEMA MG-31 Standard

While VFD basics and inverters will be discussed later in this text, there has emerged a need to provide a classification motor specifically designed for use on these electronic variable-speed drives. For various reasons (which are evident later in this text), NEMA has dedicated a section of the motor-generator (MG) specification to definite purpose inverter-fed motors. MG1-1993, Revision 1, Part 31, Section IV, is entitled "Performance Standards Applying To All Machines," Part 31, "Definite Purpose Inverter-Fed Motors." It is commonly referred to as *MG-31* and applies to squirrel-cage induction motors rated 5000 horsepower and less at 7200 volts or less for use on adjustable-voltage or adjustable-frequency controllers.

The specification addresses usual and unusual service conditions, ambient temperatures, altitudes, and ventilation requirements as well as exposure to harsh environments. Physical dimensions, tolerancing, mounting, and frame designations are covered as with standard NEMA-rated AC motors. Important items such as speed range, constant-torque operation, maximum and minimum speeds, base voltages, number of phases, rotational direction with respect to F1 and F2, and foot mounting are discussed. Service-factor and temperature-rise data are included along with data regarding the motor insulation. Insulation considerations include leakage currents, voltage spikes, shaft voltages, bearing currents, lubrication of bearings, and insulated bearings.

Operational issues such as torque, breakaway torque, breakdown torque, excess currents, and variation of rated voltages are also covered. Overspeed conditions are discussed along with sound, vibration, and torsional concerns. Nameplate marking minimum standards are set, and descriptions of motor tests, whether routine or performance, are given. Thus, the MG-31 specification does provide a fairly comprehensible guideline for applying AC motors to electronic VFDs.

Variable-Frequency-Drive Basics

Alternating current VFDs, are commonplace today in industry and in many commercial facilities. As their cost to manufacture and physical size continue to reduce, their infiltration into residential markets is also evident. Variable-frequency drives have become almost commodity items, and with good reason. A VFD can be more than just a means of changing the speed of an AC motor. It protects the motor and the electric circuit it is on with electronic current-overload functions. The VFD limits the inrush current to a motor to values far below the typical 600 percent to 700 percent that AC motors see while line starting. This makes it a true soft-start device. Thus, a VFD is a multiple-solution device for many applications. A VFD can provide the following:

1. *Energy savings on most pump and fan applications.* Any centrifugal load is an ideal candidate for utilizing a VFD.

2. *Better process control and regulation.* Setpoint control and full automated control can be achieved with VFDs.

3. *Speeding up or slowing down a machine or process.* The VFD has the ability to run a motor well past its base speed and well below also.

4. *Inherent power-factor correction.* Because the VFD is utilized in an application many times to lower the speed, a facility's power factor does not suffer as with other variable-speed devices.

5. *Bypass capability in the event of an emergency (see Figure 3–27).* The VFD can be "bypassed" and the AC motor run full speed if there is any type of problem with the VFD.

6. *Protection from overload currents.* When set up properly, the VFD protects itself and the AC motor from dangerous overload currents.

7. *Safe acceleration.* Inrush current to the AC motor is limited to no more than 100 percent as the VFD has the ability to "ramp," which is to accelerate the motor up to the desired speed. This soft start provides longer life to the motor and the machine's mechanical components.

As for nomenclature, an electronic, solid-state VFD can be called many names: motor controller; motor drive; variable-frequency drive (VFD); variable-speed drive (VSD); adjustable-speed drive (ASD); adjustable-frequency drive (AFD); the volts-per-hertz drive; frequency drive; or, in many instances, simply an inverter, which is a misnomer because there is also a converter section in the drive. Since the basis for the drive's operation is to vary the frequency to the motor in order to vary the speed, the best-suited name is the variable-frequency drive—VFD.

The major components of an electronic VFD are the power bridges and the control section. Figure 3–16 illustrates in simplified block form the two main power sections, the DC link, and the control scheme. The power bridges, the way the drive derives electric feedback from the motor, and the drive's output waveform all describe the type of drive being used. Like their DC counterpart, all VFDs have to have a power section that converts AC power into DC power. This is called the *converter* bridge and is seen in Figure 3–17. Sometimes called the *front end* of the VFD, the converter is commonly a three-phase, full-wave-diode bridge. Compared to the phase-controlled converters of older style VFDs, today's converter provides for improved power factor, better harmonic distortion back to the mains, and less sensitivity to the incoming phase sequencing.

Figure 3–16 A VFD in a block diagram.

In Figure 3–18 is shown the DC link, the *DC bus,* or *filter section.* The DC link is an important section of the drive as it provides for much of the monitoring and protection for the drive and motor circuit. It contains the base-drive fusing and precharge-capacitor network, which assures steady-voltage DC voltage levels prior to the inverter bridge and allows a path for overvoltage dissipation. This is the section of the drive's circuitry where many drive manufacturers filter the DC bus voltage. The DC voltage is monitored for surges and compared to a maximum limit to protect devices from overvoltages. Also, the DC bus can provide the "quick outlet" for braking energy to a bank of resistors whenever a motor becomes a generator. The DC bus is often called the intermediate circuit, "bus," or link.

The *inverter section* is perhaps the most important section of the VFD. As can be seen in Figure 3–19, this is the power bridge that differentiates drives. This is where that constant-voltage DC energy is *inverted* back to AC. Actually, the DC waveform looks more

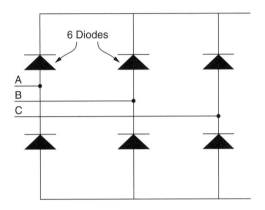

Figure 3–17   Converter of a VFD: 6 pulse.

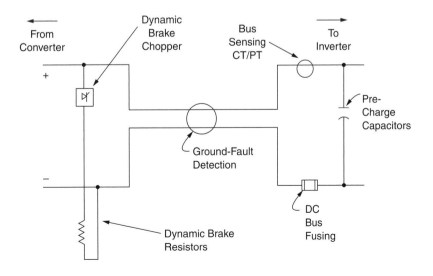

Figure 3–18   Simplified circuit showing DC bus components.

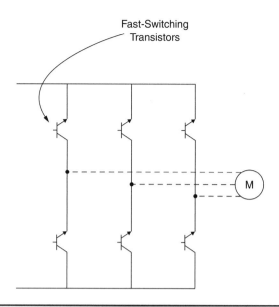

Fast-Switching
Transistors

M

Figure 3–19 Simplified inverter section of a VFD.

like an AC waveform but the voltage waveform is much different. The power semiconductors in the inverter section act as switches, switches of the DC bus, and therefore are
pulsing the motor with some voltage. The inverter, depending on the current rating of its
bridge and its particular design, can take on different shapes and sizes. Its design is so
important to the VFD that many individuals have simply started calling VFDs *inverters.*
This is not a proper name, because the drive still has other functions going on such as converting AC to DC, filtering the DC, and controlling all of these functions in addition to
the inverting of DC to AC. Some designs utilize thyristors, while most modern inverters
use some type of transistor. However, the inverter's principle of operation remains the
same—change DC energy into three channels of AC energy that an AC induction motor
can use to function properly. Inverters are classified as voltage-source, current-source, or
variable-voltage types. This has to do with the form of DC that the inverter receives from
the DC bus. It also is a function of how the drive has been designed to "correct" its own
electric feedback loop. This loop is actually part of a comparison of inverter output to the
motor with the motor's load. To continue driving the motor at the desired speed, the drive
must constantly be correcting the motor's flux.

Drive manufacturers have used varying designs of inverters: thyristor-type semiconductors, different methods of cooling, different style heat sinks, and so on. But the basic concept of the inverter has remained the same and, eventually, the thyristor has given way to
the faster-switching transistor technology. Because the thyristor has to wait for the current
passing through it to reach zero, it has become virtually obsolete in the design of inverter
sections. The transistor, able to change from conductive to nonconductive almost instantaneously, has become the device of choice. It can be made conductive with only a 3- to 5-
volt source of power, thus making it very efficient and keeping heat losses to a minimum.
Today, switching frequencies are in the 15- to 16-kilohertz range for most inverters on the
market today.

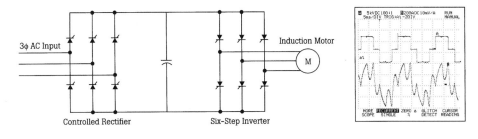

Figure 3–20   Voltage-source inverter, also called six-step. Note voltage and current waveforms (Reproduced with permission from Fluke Corporation).

Variable-Frequency-Drive Types

Solid-state AC VFDs can be named for their use, by their DC bus/inverter voltage or current source, by their waveform (PWM or PAM), by the type of power device used in their inverter section, or even by their performance characteristics. By use, there are AC traction drives, vector drives, load-commutating inverters, spindle drives, sensorless vector, volts per hertz, and many others. When classifying a VFD by its supply of voltage or current to the inverter, we get variable-voltage inverters (VVI), voltage-source inverters (VSI), and current-source inverters (CSI). When the name describes the drive's output waveform, VFDs can be called pulse-width modulated (PWM) or pulse-amplitude modulated (PAM). Lastly, we have VFDs being referred to as transistorized, IGBT (insulated-gate-bipolar-transistor) type, or even six-step-SCR types, which describe the inverter devices being implemented.

Variable-frequency drives are usually classified by their use of the DC bus, the drive's output, or the shape of that output's waveform. The main objective of the VFD is to vary the speed of the motor while providing the closest approximation to a sine wave for current (while pulsing DC voltage to the motor). After all, when an AC motor runs directly off of 60-hertz power, the signal to the motor is a sine wave (as clean as the local utility can provide). Put a variable-speed drive in the circuit and vary the frequency to get the desired speed. Sounds simple enough, but the industry is continually striving to address all the side effects and to provide a pure system.

Voltage-Source Inverters

If the inverter receives a constant DC voltage via the DC bus, then it is said to be a *voltage-source inverter* (VSI). As is seen in Figure 3–20, the inverter has to operate to control both the switching on and the switching off of that bus's DC voltage; the corresponding waveforms are also shown. The DC bus power must be constant and clean all the while that the proper frequency of pulses is maintained for the given speed. Pulse-width-modulated drives are voltage-source inverters.

Variable-Voltage Inverters

If an inverter receives a DC voltage that varies, then it is called a *variable-voltage inverter* (VVI). Some versions are called *six-step drives*. They use different devices in the power bridge, and, as the name implies, there is a step in the voltage whenever the devices turn on and off. In this case, as the voltage is variable, there are unsteady current waveforms that are undesirable.

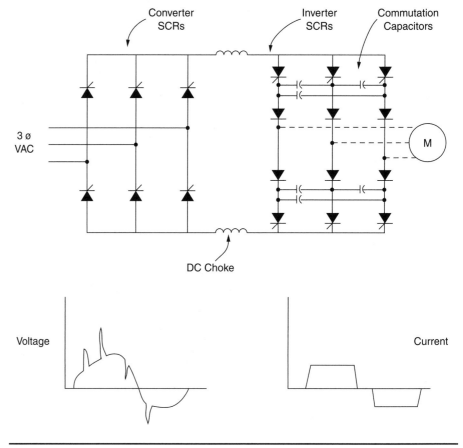

Figure 3–21 Simplified current-source drive circuit with voltage and current waveforms shown.

Current-Source Inverters

Still another type of inverter is the one that sources a DC from the DC bus. Sometimes this drive is called a *CSI,* which is short for *current-source inverter.* The equivalent circuit and waveforms for this inverter are shown in Figure 3–21. Note the magnitude of the commutation notches (these notches make it hard to meet IEEE519 harmonic guidelines, as will be seen later). This inverter will normally utilize SCRs as switches to gain a six-step-current-waveform output. Here, the conducting time is changed up or down for each individual step, resulting in a longer or shorter cycle time. The voltage waveform looks somewhat sinusoidal, with the exception of the commutation notches, which occur every time an SCR turns on and off.

Pulse-Width Modulation and Pulse-Amplitude Modulation

The PWM VFD is the most common drive design used today. It integrates fast-switching transistors into the inverter section to accomplish the switching pattern. The switching pattern is to control the width of the pulses out to the motor. The current and voltage waveforms are shown in Figure 3–22. The output frequency of a PWM drive is controlled by applying positive pulses in one half period and negative pulses in the next half period. The DC voltage is provided by an uncontrolled diode rectifier. Thus, by switching the inverter-

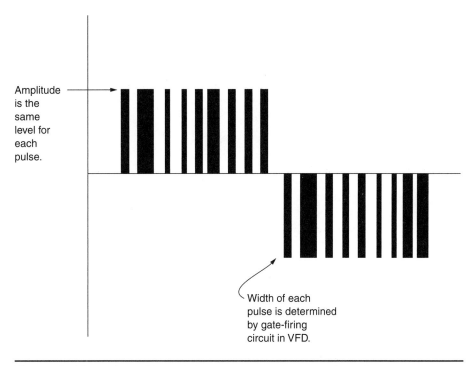

Amplitude is the same level for each pulse.

Width of each pulse is determined by gate-firing circuit in VFD.

Figure 3–22 Pulse-width modulation (PWM). The output from the VFD inverter to the motor.

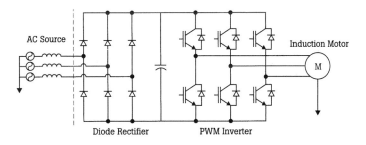

AC Source

Induction Motor

M

Diode Rectifier   PWM Inverter

Figure 3–23 Simplified circuit of the PWM drive (Reproduced with permission from Fluke Corporation).

transistor devices on and off many times per half cycle, a pseudosinusoidal current waveform is approximated. An equivalent circuit with several key components highlighted for a PWM VFD is shown in Figure 3–23.

The PAM VFD is more concerned with the amplitude of the pulse than with the frequency. It is a variable-voltage-type inverter because the value of the voltage varies as it is being switched. This waveform is typical for VVI designs. Whereas a PWM drive and its high-switching frequency may affect audible motor noise, the PAM drive can also have some adverse effects on the motor. These may include cogging (erratic motor rotation) at low speeds and increased heating due to voltage spikes in the waveform. The PAM-drive controllers are not that common at this time.

The more common designs for the converter utilize diodes (most) or thyristors (some) for rectifying the AC incoming voltage into DC voltage. In a diode rectifier, the three-phase, incoming, 60-hertz power is channeled into three legs of the converter circuit, each leg with two diodes. This creates a constant DC voltage, 800 volts DC maximum for 460-AC-volt-supplied units and 400 volts DC maximum for 230-AC-volt-supplied units. From here, this constant DC voltage goes through the DC link to the inverter circuit to be pulsed to the motor. The diode rectifier is the most popular design because it is simple and the least expensive. Advantages of this type of rectifier include unity power factor, less distortion backfed to the supply, and resilience to noise in the converter itself. The biggest drawback to using diodes is the fact that no voltage can return to the source, therefore allowing no regeneration. Another separate power bridge must be added to the converter section if regeneration is required, and this can be expensive.

The SCR converter has more capabilities than the diode converter because it is more controllable. However, for VFD considerations, the more important controlling section is the transistorized inverter. The SCR bridge is more robust and flexible than the diode bridge but will cost more. Perhaps the biggest disadvantage of the SCR, versus the diode bridge, is its voltage and current distortion contribution to an electric system. Silicon-controlled-rectifier commutation notches create dramatic disturbances on the supply and simply cannot be tolerated in facilities. As SCR designs have been employed, their design is a full-wave-type rectifier. Six SCRs are used to control the gating of the device. *Gating* is the term given to controlling the SCR's time of conduction by turning the SCR on and turning the SCR off. The SCR cannot be turned on until it has deenergized after being commanded to turn off. This is sometimes referred to as the *zero crossing* of the current. Thus, many SCRs have different turn-off times and it is sometimes necessary to get all six SCRs to match in one drive's circuit to assure proper, smooth gating. The drive's logic circuitry provides the means of control for this gating sequence, thus controlling the output voltage to the inverter. Additionally, with SCR designs, as speed is decreased, so too does the power factor of the system. This is a nonattractive, inefficient use of voltage and current within a facility, which can promote penalties from the utilities. Another issue, mentioned earlier, is that of distortion fed back to the supply—a major concern. A choke or otherwise special circuit filter often must be added to minimize disturbances, and this drives the cost of the overall drive system up. Additionally, these SCRs are more susceptible to line disturbances themselves, which can result in nuisance drive tripping. As long as manufacturers take measures to protect against these problems, this type of rectifier is attractive, especially because it has the ability to regenerate power back to the AC supply simply by gating the SCRs in reverse sequence.

The latest technological design method for inverters is based around the insulated-gate-bipolar transistor (IGBT). This transistor is a combination of features provided by the MOSFET and the bipolar transistor. It has good current conductance with lower losses. It possesses very high switching frequency, fast rise times, and is easy to control. It takes only 3 to 5 volts of energy to make it a conductor. This technology has gained much momentum as the IGBT can be used in horsepowers up to several hundred horsepower, thus allowing higher current capacities.

Many inverter designs utilize high-switching-frequency transistors such as MOSFETs, bipolar transistors, or Darlington transistors. The transistor's basic advantage is that it can be switched from a conducting to nonconducting state extremely fast. It does not have to wait for a zero-crossing condition like its diode and SCR counterparts. Also, since higher current ratings of transistors are now available, higher resulting horsepower drives are being built, sometimes with transistors in parallel to get the desired output to the motor. These transistors have the ability to switch at several kilohertz. This virtually eliminates audible noise at the motor, which was an earlier objection to using high-switching-frequency transistors. This also means that the current waveform to the motor is close to sinusoidal, meaning smoother motor operation especially at low speeds.

How Variable-Frequency Drives Operate

Not only does the VFD control the AC motor, but it also becomes part of the motor's electric circuit. The drive needs the motor, and the motor needs the drive. The proper operation of the motor/drive system is critical to each supplying the other with voltage and/or current, and vice versa. As was seen earlier in the following formula for an AC motor:

$$Synchronous\ speed = \frac{120 \times frequency}{number\ of\ motor\ poles}$$

The value *120* is a constant and cannot be changed. It is derived from the electric relationship given to synchronous machines with fixed poles, location of those poles in a given half cycle, and the frequency (cycles per second and seconds per minute). Thus, for a given motor, the number of poles has to be constant; they are physically in place on the motor's rotor. Therefore, to change the speed, all that can be changed in the formula is the frequency, which is exactly what a VFD does. In the equation, if the frequency is 60 for normal 60-hertz supplied power, then the motor speed will be the maximum for an equating number of poles. This is referred to as the *synchronous speed.* As an example, 60 times 120 is 7200 and goes in the numerator. A two-pole motor has just that—two magnetic poles, north and south. Therefore, 7200 divided by 2 equals 3600; thus, this is the speed in rpm that that motor will run at if it is applied 60-hertz power. Likewise, if the frequency is 30 hertz, then the resulting speed is half, or 1800 rpm.

As was shown earlier, the number of poles in the AC induction motor is fixed. These magnetic regions on the motor's rotating element, the rotor, have a permanent polarity, plus or minus. As the stator windings receive electric energy from the VFD, a change in magnetism takes place; thus, motion of the rotor takes place. The *frequency* is how often this current flows through the windings. It should be noted that when a VFD first applies power to the AC motor, a certain amount of current is need to get magnetic flux built up enough to even begin motion. This is called *magnetizing current* and must be accounted for by the drive before any torque-producing current can be utilized. Magnetizing current represents approximately 25 percent to 30 percent of a motor's total full-load current. When driving an actual load, the actual shaft speed of an AC squirrel-cage motor will be slower than the synchronous speed. This is called the *full-load speed* and is a function of a motor characteristic called *slip* and is typical of all AC induction motors. Because the motor is always dynamically correcting to maintain speed, when loaded, it lags behind in actual motor rpm. As a review, the percentage of slip can be found by using the following formula:

$$Slip\ percentage\ = \frac{synchronous\text{-}motor\ rpm - full\text{-}load\ rpm}{synchronous\text{-}motor\ rpm} \times 100$$

With respect to the VFD, motor slip is going to directly relate to the ability to drive any given load at a given speed. Torque can be said to equate to load, which equates to current. Therefore, there are given relationships to speed and torque. Slip will remain constant anywhere on the speed/torque curve as the frequency is reduced to achieve the desired speed.

Slip is a critical element in controlling an AC motor, especially at low speeds. In essence, controlling slip means the motor is under control. Alternating current variable-speed drives with volts-per-hertz capability can control slip very well down to low speeds. The biggest factor is the loading. Light or even centrifugal loads are much easier to control at slow speeds.

Torque is produced as the induction motor generates flux in its rotating field. This flux must remain constant to produce full-load torque. This is most important when running the motor at less than full speed. And since VFDs are used to provide slower running speeds, there must be a means of maintaining a constant flux in the air gap of the motor. This method of speed and flux control is called the constant volts-per-hertz ratio method of operation. It is typical of most PWM, solid-state VFDs. The PWM drive's ability to maintain the necessary AC levels through all types of load conditions at given speeds is that factor which separates one drive manufacturer from another.

Volts-per-Hertz Control

When changing the frequency for speed control, the output voltage changes proportionally with respect to the curve in Figure 3–24. The volts-per-hertz ratio is a linear function and nominally is 7.67 to 1 for a 460-volt, 60-hertz system ($\frac{460}{60} = 7.6$). Thus, at half speed on a 460-volt supplied system, the frequency is 30 hertz and the corresponding voltage is 230 volts to the motor from the drive. Likewise, on a 230-volt supplied system, the volts-per-hertz ratio is 3.83 to 1. Thus, at half speed, 30 hertz, the output voltage should be 115 volts. Measuring these values with metering equipment is sometimes an adventure. Depending on the type of meter used (analog, true rms, average, or even a scopemeter) and how fast or slow it samples data, the readings can be all over the place.

Even though the output voltage, based on the programmed volts-per-hertz curve, shows a lower value than full voltage, the inverter is still pulsing continuous higher levels of DC voltage out to the motor, which is enough to fully excite the motor windings and continue operation. A 460-volt drive is typically pulsing at 650 volts DC, and a 230-volt drive is pulsing at 325 volts DC. Thus, whenever the output voltage reads half of rated, there is still an average power value, which would equate as is shown in all the pulses of energy in Figure 3–25. If sine wave power could be delivered to an induction motor with an rms level (Figure 3–26) equal to a lower voltage for the lower speed, then the average power is shown

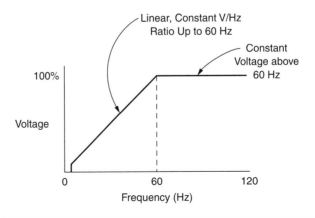

Figure 3–24 The volts-per-hertz (V/Hz) operation of most VFDs.

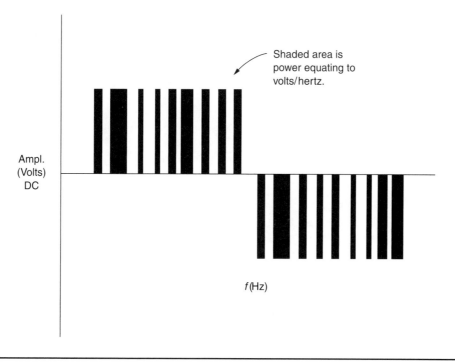

Figure 3–25   The area within each pulse is the power delivered to the motor in volt-microseconds.

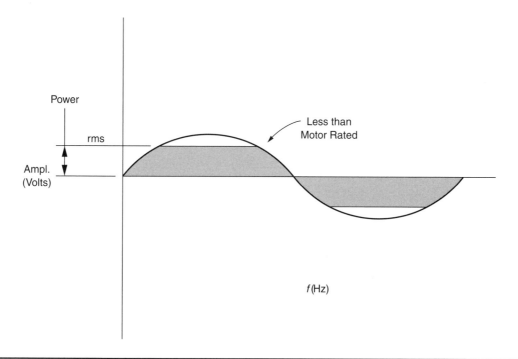

Figure 3–26   *If* a sine wave, with an rms amplitude of less than motor rated voltage, could be used.

as the area under the curve in terms of volt-microseconds. The PWM pattern makes the VFD save energy to the motor, but it is also very critical to performance. The VFD tries to maintain this ratio, because, if the ratio increases or decreases as motor speed changes, then motor current can become unstable and torque can diminish. This is the reason that VFDs start to have control troubles below 10 hertz on constant-torque loads. The flux vector design VFD is one solution for maintaining better control of tougher applications at very low speeds. Another method of increasing the voltage at low speeds to produce adequate torque is by incorporating a voltage-boost function available on most drives. However, if the motor is lightly loaded and voltage boost is enabled at low speeds, then an unstable, growling motor may be the result. Voltage boost should be used when loading is high and when the motor must run at low speeds or start with a high load.

Alternating current VFDs also allow for motor operation into an extended speed range, sometimes called overspeeding or overfrequency operation. This is also known as the constant-voltage mode of operation. Sometimes the application requires that the motor run beyond 60 hertz. Frequencies of 120 (twice base speed), 200, and 500 hertz, and even beyond, are possible with faster-switching inverters. Higher speeds can be achieved, but torque diminishes rapidly as the speed goes higher. Trying to maintain a constant ratio between the voltage and the frequency from zero to full speed is desirable in these kinds of applications.

Where this capability really is a benefit is in those pumping applications where getting a little more speed out of the motor can increase the flow just enough to satisfy the demand. When there is capacity in the motor and the VFD can be programmed to do this, a new, larger motor does not have to be purchased and installed. The old motor does not have to be taken out. Often times, the VFD can run 10 percent to 20 percent higher (66 to 72 hertz) in speed and make up for lost capacity in a flow-and-demand type of system. High-speed applications include test stands and dynamometers. While this is an attractive feature of the VFD, care must be applied when utilizing this function. Many applications cannot tolerate going extremely fast. Many motors, as built, are not balanced for these high speeds. Physical components such as gear boxes, couplings, fan blades, and so on can actually explode at too high a speed. This can be a dangerous situation. However, many test stands are actually rotating an object as fast as possible to see when it does break apart. For safety sake, investigate the mechanical system accordingly before experimenting with high speed operation from a VFD.

Variable-Frequency-Drive Selection

Selecting a VFD for an application is not always clear-cut. There are tough questions to ask, and, usually, there is more than one answer for each question. Each application's needs dictate what type of drive is required. Do motors exist and what type are they? Should we use a DC motor or an AC motor? Once that issue has been decided, then the application requirements take over. What are the torque requirements of the application—constant, variable, high starting? Will there need to be braking or regeneration? What is the horsepower? What is the supply voltage? What speed regulation will the application require? Is torque regulation more important? Is a digital style or analog drive available? What will the maintenance person find when there is a problem? What kind of duty cycle or loading is predicted? Does the plant have a "clean" voltage supply? The questions seem to be endless, but keep asking, because the more that is known and shared about the application, the more likely it will be successful.

Considerations

One drive manufacturer's VFD may be better suited for one application than another. There are hundreds of considerations to ponder when selecting a VFD. Section 16, Electrical, of Architectural Specifications usually contains written descriptions of drives to be used on building and construction projects, but this version is usually too generic. The Institute of Electrical and Electronics Engineers (IEEE) has a good drive-specification standard that is more detailed for drive usage. Most drive manufacturers can also provide specifications. A copy should be secured when purchasing or just to get a better understanding. Technical vendors—as many as possible—should also be consulted, because the industry is constantly changing. They can advise if the application is right for their drive and vice versa. Following is a list of issues and possible questions to ask when VFDs are involved.

Location: Where will the drive be physically located? What are the ambient conditions? How far from the AC motor will the drive be located? Just as with real estate, a VFD's successful implementation depends on its location, location, location!

Constant Torque or Variable Torque (CT versus VT):   What does the load demand? How much current is required on a continuous basis for the particular application? What does the speed-versus-torque curve look like? Perhaps the application only needs starting or peak torque for a few seconds; if so, then one frame size of a VFD may be more suited than another (it may have to be upsized to handle the starting current).

Speed Range or Turn-Down Ratio: What will be the top speed needed and the lowest speed needed? What is the loading like at these various speeds? Turn-down ratios of 15:1 and greater are common for variable-torque applications. Low speeds while fully loaded are the most difficult for VFDs to control. Similarly, hard-to-start loads may require high breakaway torque. Typically, a drive's slip-compensation function can provide adequate adjustment. Additionally, low speeds often mean a need for an auxiliary blower at the motor. When a motor runs at low speeds, fully loaded, for long periods of time, it must be fitted with an external blower with a small motor. Also, the starter for this blower motor, often just a single-phase device, must be included within the drive package. This auxiliary blower will run constantly, regardless of the motor's speed, and blow air continuously over the skin of the motor.

Input Power Supply: Some drives are more sensitive than others. Some control circuits of VFDs can tolerate 5 percent dips while others can handle 10 percent. Also, determine prior to the drive installation if the power supply will be stable. If the supply has voltage dips, frequent outages, and surges, then take necessary precautions at the drive. Undervoltage trips will occur and become a nuisance. Constant-voltage sources and transformers can provide some solutions.

Wire and Cable: Motor voltage is directly proportional to speed in a VFD variable-speed application. Voltage drops can produce reductions in torque and are noticeable especially at low speeds. When the motor is running near full speed, a voltage drop in the wire does not really affect the speed. Additionally, special cable is available for use between the drive and the motor, which minimizes overvoltage spikes. The routing and the size and type of cable used can be very beneficial to the VFD/motor installation. *The closer the VFD is to the motor, the better.*

Grounding: Variable-frequency-drive and motor applications requiring three-phase power need to use four-conductor cable. Additionally, there are many types of "VFD cable" available on the market. Some have unique methods of handling overvoltage spikes, excess cable heating, and grounding. These should be fully investigated and applied in a VFD installation if necessary. All equipment in the motor/drive circuit must be tied to earth ground at one location. The input power to the drive should be from a wye-configured source. All current will be contained within these four wires and will minimize interference on the input. The output wiring should allow for the fourth wire, the ground conductor, to be used as the fixed-ground connection between the VFD's enclosure and the motor itself. *National Electrical Code®* Article 250, Grounding, should be consulted whenever issues arise concerning ground systems with VFDs and motors.

Replacing a DC Drive with a VFD: Check the running, accelerating, and starting torque requirements. An identical horsepower AC motor and drive may not provide the necessary torque. A DC motor can often be run into the field-weakened speed range, thus keeping its horsepower size down. A VFD and motor may have to be upsized in order to get the proper torque at all operating speeds. Applications that may involve this are center winders, mechanical-pulley drives, eddy current clutches, and other constant-torque loads.

Multiple Motors Operating off of a Single VFD's Output: A PWM, voltage-source drive is more suitable here. One important issue, though, is to apply individual motor thermal-overload protectors. The VFD could inadvertently supply too much current to a smaller motor because the drive does not necessarily know it is running smaller motors. The drive's electronic-overload protection is normally for a motor sized for the drive's rating. Supply and return fans are sometimes both operated from the same VFD as they tend to run the same speeds anyway.

Power Factor: Most of today's PWM drives have a unity power factor in the 0. 98 range. Power factor must be taken into account due to utility penalties. The utility imposes a penalty on demand charges for a low power factor. A drive with a constant power factor throughout its speed range is attractive, especially when it will replace a drive that has a poor power factor.

Single Phasing into a Three-Phase Drive: Variable-frequency drives, are phase-sequence insensitive; they only want that DC bus charged. Therefore, single-phase input to a drive running a three-phase motor is something that can be done. However, the drive will have to be derated for the desired output at the selected motor. Typically, this derating is by 33 percent of full drive output capacity. For example, a $\frac{3}{4}$-horsepower drive with single-phase input will be derated to a $\frac{1}{2}$-horsepower drive $\left(\frac{3}{4} \times \frac{2}{3} = \frac{6}{12} = \frac{1}{2} \right)$.

Carrier or PWM or Switching Frequency: Named as the frequency that "carries" information, this wave carries pulses to the motor for operation. Today's PWM drives switch at 15 to 16 kilohertz, whereas years ago the maximum was 2.5 kilohertz. The more often the transistors turn on, the more often current conducts—thus, more heating. The value of the carrier frequency in any VFD in service should be known. Lowering or changing that value can have definite results depending on the installation. Consult the drive manufacturer concerning this.

Variable-Frequency-Drive Efficiency: A VFD's input-current rating is different than its output-current rating. There are losses due to heat within the drive. Efficiencies at all speeds—and mainly for the predicted speeds to be operated—need to be considered. Remember, the VFD is supposed to be an energy-saving device by its motor control, but its own hardware design should also be energy efficient. Pulse-width-modulated-drive efficiencies can be taken right from the drive nameplate: current out/current in.

Ventilation and Cooling of VFDs: Variable-frequency drives can be air-cooled, water-cooled, or air-conditioned and can dissipate heat through a surface area. Air-cooled drives tend to be noisier than the other types due to the cooling fans inherent to the package. Water-cooled-drive systems are more involved. They need a pump, piping, and a heat-exchanger system in addition to the drive's converter and inverter bridges. These water-cooled systems, although very efficient when maintained, have the potential for leaks, must be cleaned, and have higher initial costs.

When physically installing any drive, whether it is AC or DC, special attention should be given to the heat generated by the drive. It has current running through it that produces heat, and this heat has to go somewhere. It can naturally dissipate if there is a light-duty cycle. But, if loading is heavy, then provisions for cooling or ventilating are necessary. First, take a look at the ambient environment around the drive. Is the room in which the drive will be located warm or hot naturally? A VFD typically is specified to run properly in an ambient temperature of no more than 40°C. What will be the temperature on the hottest summer day? Determine if the drive is going to be heavily loaded. Will it run 24 hours a day or intermittently? Most of the time, ventilation fans of sufficient size, pulling ambient air into the enclosure and up across the drive, will be adequate. However, if dirty air is brought into the enclosure, then a new problem can emerge. Dust and dirt will collect on the drive and virtually suffocate it. Eventually, no heat will be able to escape from the drive and it will overheat. Most newer drives will trip or fault on an overtemperature fault. This protects the drive but is a nuisance because the drive will have to cool off before it can be started again.

Two other methods of handling drive-enclosure heat are by air-conditioning the cabinet or by providing a heat-exchanger system. Both are more expensive approaches but sometimes the only answer. When the ambient air is too warm or too dirty, then air-conditioning makes sense. There are many manufacturers that specialize in small, compact air conditioners that attach directly to the wall of a drive enclosure. These are self-contained, closed-loop units that keep the inside of the drive enclosure completely cooled. Additionally, if the ambient area is dirty, then the air conditioner will have to be kept clean to operate efficiently. Also, if the air conditioner stops running for whatever reason, the drives will trip quickly as they are now in a completely sealed enclosure with no way for heat to escape.

One other cooling system often employed with VFDs is the vortex cooling system. By bringing pressurized-compressed air to a drive enclosure, the cabinet can be kept both cool and under positive pressure. Positive pressure means that dirt will not seek a path into the enclosure. The pressurized air must be clean and dry before entering the drive enclosure. Often a filter must be utilized.

High Altitude and High Humidity: These are not the most common issues when it comes to drive selection, but certain geographic areas around the world present a different set of circumstances. High-altitude VFD installations can pose a problem as the altitude increases; so, too, does that air's inability to dissipate heat. Air at 3300 feet above sea level (or approximately 1000 meters) cannot hold as much heat as an equivalent amount of air at sea level. Therefore, transferring the heat from the heat source, the drive, is more difficult at higher altitudes. Drives often have to be derated for every 1000 feet above the specified 3300 feet above sea level.

Likewise, high humidity and excessive-moisture installations also pose problems. A problem arises for a drive if the excessive moisture in the air condenses on the drive's components, especially when it heats and cools. Water will conduct electricity, and, if enough water forms on the drive, short circuits can occur. Most VFD specifications allow a maximum level of relative humidity of 95 percent to 96 percent, but noncondensing. This problem can cause actual board damage and can cause nuisance tripping. One means of dealing with condensation is to install space heaters within the drive enclosure. These should come on when the drive is off. While the drive is running, enough heat is generated by the drive itself to keep moisture from condensing. Once the drive is shut down, the heat production will also cease. The circuitry should be such that the space heaters should then energize. Some installations can even make use of a standard lightbulb as the space heater.

Input and/or Output Reactors: As was discussed in Chapter 2, coiled inductors—called *reactors*—can add impedance to a VFD system and have a positive effect. To minimize noise in and out of the drive at the supply point due to the rectification of the diodes, an input reactor is often installed. It also helps to suppress transients from entering the drive. The input reactor is typically less expensive than the drive-isolation transformer (DIT), but there are some basic rules for selecting one over the other: (1) If the AC supply does *not* have a neutral or one phase referenced to ground, then the DIT is recommended. (2) Whenever the kilovolt-ampere capacity of the AC supply is four times the horsepower rating of the VFD/motor, either can be used. (3) If there are power-factor-correction capacitors being switched on and off in the circuit, then 5 percent reactors or DITs are recommended to go between the VFD and the capacitor bank. (4) If line-to-ground voltages can exceed 125 percent of nominal (on any phase), then a DIT is required. (5) If voltage spikes and transients are present in the system, then either the DIT or the 5 percent reactor should be installed. A reactor is sometimes needed to reduce ringing at the output to the motor; it slopes off the fast-rise time of the transistors and minimizes overvoltage spikes to the motor. Another reason to use output reactors is to protect the inverter power devices from a short circuit on the drive's output.

Adjustable Variable-Frequency-Drive Functions and Features

Most VFDs today, being microprocessor based, have many built-in features. The true capability of the VFD is not only in the converting and inverting power sections but also in the software that allows the user to gain full control of the drive and motor installation. Earlier versions of VFDs were analog and had to include extra, printed circuit boards or plug-in modules to accomplish many specialized tasks now included in a VFD's standard parameter list. With digital technology, high-speed microprocessors, and ample memory, all one has to do is call up the parameter on the drive's display and make the necessary change.

Set-up Parameters

Setting up a VFD is easier and quicker today than in the past. Commonly used set-up parameters for VFDs include the following possibilities.

The Control Method: Since many VFD installations do not get the control scheme "designed" until the day of commissioning the drive, there has to be a lot of flexibility built into the VFD. Luckily, the start, stop, reference-signal type, stop method, remote control, or local control functions are all in the VFD and can be programmed for a wide variety of operator/engineer/owner needs and wishes. If not built in, then a lot of relays would have to be wired in!

Acceleration or Accel-Ramp Rate: This function allows the user to select the amount of time desired to reach full speed. Often referred to as *ramping,* this function actually provides the soft-start capability that limits inrush current to an AC motor. High-inertia loads, such as large blowers or fans, may need several seconds, even minutes, to accelerate to full speed. If this were not adjustable, then the drive would constantly trip off line, trying to supply too much current too fast (overcurrent fault) under these conditions. Some drives can provide two, three, and even four different acceleration settings. As the drive begins to accelerate the motor, a contact closure can signal that it is time to "shift gears" and go into a faster ramp up to speed. Note that the acceleration rate is "at work" even when any speed change happens, however small it may be. Going from low to high speed will utilize the set acceleration rate. S-curves or S-ramps provide for real soft starts. The premise here is that an extremely soft start is needed, but speed and time have to be made up in the mid-portion of the curve, before softly settling into the set speed (see Figure 3–27).

Deceleration or Decel-Ramp Rate: Similar to acceleration in setup, deceleration most often has different value and is used to provide controlled stop of the motor. It has limits

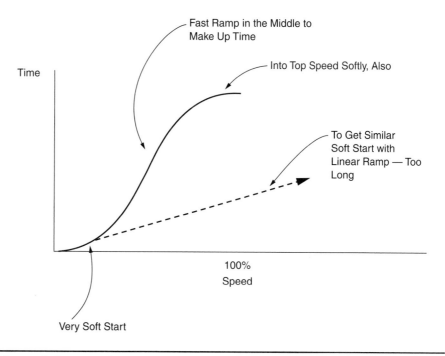

Figure 3–27 The S-curve acceleration ramp.

and is load dependent. If a high-inertia load is to be brought to a fast stop, then the drive must have some method of handling the motor-generated energy—the overvoltage condition. The drive can provide some internal braking power by allowing its DC bus voltage to rise slowly. If this voltage rises to the maximum set level (set there to protect devices), then the drive will trip on an overvoltage fault. The solution is to try a longer deceleration rate.

Automatic Restart: Most VFDs can automatically attempt to restart themselves when conditions permit; however, the VFD installer must determine if it is safe to have the motor automatically started. A drive can be programmed to automatically attempt to restart up to ten times at two-minute intervals. As long as power is available to the control circuitry, the drive's logic unit will allow the restart attempts. If, after the select number of attempts, the drive cannot restart, then it remains in a faulted condition until someone manually resets it (and the cause is determined). A typical automatic-restart function is appropriate whenever supply power is known to dip below acceptable levels of voltage (undervoltage).

Stopping Method: Variable-frequency drives can be programmed to stop in a variety of ways. They can allow the motor to coast to a stop; they can ramp or decel to a stop; they can employ DC injection braking (which applies a DC voltage to one phase out to the motor to produce an opposing force to brake it); or a mechanical brake can be commanded "on" in an E-stop condition.

Automatic Signals: Many drives can be programmed to run manually from a potentiometer or keypad setting right at the drive. Other times, it is desirable to receive a 0 to 5 volt, 0 to 10 volt, or 4 to 20 milliampere signal to scale as the speed range. Four milliamperes will equal minimum speed, and 20 milliamperes will equal maximum speed. These minimum and maximum values are also programmable for the particular application. When running in the automatic mode, it is necessary to have a safety built in that handles those conditions where the automatic reference is zero (absent) or loaded with electric noise and outside the minimum/maximum settings. Most often, the drive will go to a preset, safe, slow speed under these circumstances rather than to allow the motor to run away. Some drives can be programmed to also shut down and fault, while other applications may need to have the motor keep running while a fault is announced. Likewise, many VFDs, which may be accepting an encoder or tachometer feedback signal, may need to fall back to a safe, slow speed if that signal is lost.

Jump, Skip, or Critical Frequencies: In many drive applications, especially fan and pump applications, there is always the possibility that resonant frequencies can exist. Many times, these resonant frequencies can cause severe vibration in the mechanical drivetrain of the drive-and-motor system. If the drive were set at this frequency and run continuously there, then possible premature mechanical failure could occur. The VFD can be programmed to avoid these certain frequencies by selecting a frequency above or below (with a set bandwidth) in order to skip over the known resonant frequency.

Fault Logs and On-Board Diagnostics: Many digital drives can display the fault that has taken them out of service. Many will log the last four, eight, or more faults and list what they are. This is a good troubleshooting tool, especially to detect trends. Some faults automatically restart and keep the motor running, and the plant personnel are not aware of any problems. Some VFDs have enough memory to store faults as they occur and thus have a record of when a fault happened, what it was, and how the drive was reset (automatically or manually). Typically, these faults are stored in a first-in, first-out manner.

Power Loss Ride Through: A VFD has the ability to ride through short power interruptions. The VFD will continue to run off of stored DC bus energy until the bus voltage reaches a level where it can no longer supply enough voltage to keep running, and it then faults. Most drives can only ride through 1 to 2 seconds.

Slip Compensation: The amount of slip in an AC motor system is proportional to load. With increased slip comes an increase of necessary torque to continue driving the load. However, many times the speed of the rotor actually slows in order to continue driving the load. In applications where this speed shedding is undesirable, slip compensation provides for a solution. As the load increases, the VFD can automatically increase the output frequency to continue providing motor slip without a decrease in speed. The actual amount of slip compensation will be proportional to the increase in load. Usually one setting of slip compensation will cover the entire operating range with speed regulations of around 0.6 percent.

Catch a Rotating Motor, Speed Search, or Pick Up a Spinning Load: Many VFDs are expected to catch a motor spinning with load at some speed and direction and then take it to the actual commanded speed and direction. This means that the drive has to be able to determine the present motor speed, instantaneously, determine the direction, and then begin its output at that speed. The drive then has to reaccelerate or decelerate the motor to the desired reference speed. Applications with large, high-inertia fans are good applications for this function. A large fan blade can take several minutes to come to a rest after running full speed. The VFD can start right into this coasting motor.

Motor Stall Prevention: There are some instances where a motor can get into a temporary overload condition (current limit). The drive's normal operation wants to protect the motor and itself and thus shut down or trip. This can be a nuisance fault and may not be acceptable. Many fan systems that begin moving cold air experience these overload conditions. But, once the air warms, the motor becomes less loaded and will continue to operate as required. This function allows the drive to lower the output frequency until the output-current levels begin to decrease. In this way the drive will ride through the overload condition without tripping.

Variable-Frequency-Drive, Constant-Speed Bypass System

Should the VFD fail or need maintenance, the actual drive hardware can be completely bypassed and the AC motor can be run full speed. As can be seen in Figure 3–28, three interlocked contacts can completely isolate the VFD from the 60-hertz supply. When operating properly, the VFD line and load contacts are used, but, when there is a problem with the drive, the bypass contact closes and the line and load contacts open. This can be done automatically whenever the drive trips or can be done manually. It must be noted that, when in bypass, the motor will go to full speed as fast as it can. This condition may not be suitable for existing ductwork; damage to the duct system can occur. It must be verified that dampers are open and this condition is allowed.

High- Performance, Specialty, Variable- Frequency Drives

There has evolved a class of specialty, electronic VFDs that are called high-performance drives. An application often dictates that performance—speed and torque regulation of the motor—outweighs energy savings and all other beneficial factors that VFDs provide. Therefore, taking the basic VFD design, a faster-responding and better-controlling device is made; and they are called different names by different drive manufacturers. This group of drives is quickly approaching the performance criteria of complex servo systems, but for

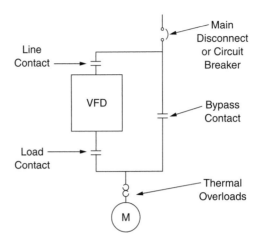

Figure 3–28 A full-voltage bypass, single-line diagram.

less cost. While the flux-vector drive is one type of high-performance VFD, its open-loop cousin is the sensorless-flux vector; another is the field-oriented controller, and still another is named the direct-torque controller. It is often hard to actually see much real difference in flux-vector, field-oriented, and direct-torque-control designs. To the drive-design engineer, it is obvious; to the plant operator—"Just run my machine precisely and keep me in production!" However, there are some real differences.

Flux-Vector Control (FVC), Including Sensorless-Vector Control (SVC)

The vector control principle states that the magnitude and phase of the AC motor's stator-current vector is to be controlled by producing constant magnetic flux while generating the necessary component of torque-producing current. This high-performance VFD is called the *flux-vector drive*. It takes complete control of the motor and load by using special algorithms, high-speed microprocessing, and digital feedback from the motor itself. Often times, an AC flux-vector system (motor and drive controller) can outperform an equivalent horsepower DC system from a speed and torque regulation performance standpoint.

Figure 3–29 shows the vector relationship to the torque angle. The torque angle mathematically is the arc tangent of the torque vector divided by the magnetizing vector. It is shown as it relates to the magnetizing current, I_m. Keep in mind that these events are taking place within the air gap of the induction motor. Thus, physically within the motor, the torque-producing current is trying through all load changes to remain 90 degrees from the magnetizing current. The magnetizing-current vector is continuous and completely independent of the torque vector. There can often be discontinuous current in DC drives. This makes the continuous-current feature of the AC flux-vector drive attractive.

The flux vector drive has incorporated the basic AC power bridge and the converter and inverter sections and added some extra control algorithms, as is seen in Figure 3–30. Thus, any design of VFD could be modified into a flux-vector version. Pulse-width-modulated drives, current-source drives, and others have been made into vector packages. The power and performance of the individual vector design is in the control circuitry, software, and integrity of the feedback device. The direct access to the torque-producing vector even allows the vector drive to function as a torque-reference follower, or torque helper. A torque-reference signal can be sent from the drive, or the drive can receive a torque command.

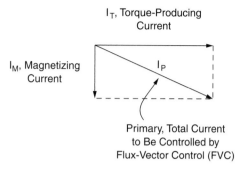

Figure 3–29 The vector relationship in a flux-vector-control scheme.

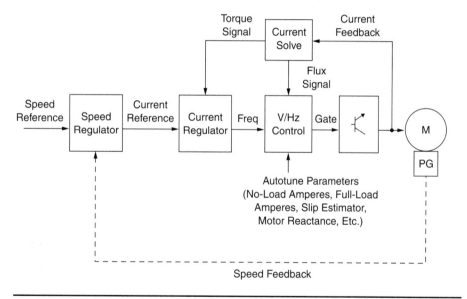

Figure 3–30 A flux-vector-drive circuit.

By attaching a pulse generator to the motor's rotor and feeding this signal back to the drive, motor-slip information and rotor position can be determined. This feedback signal is critical to the flux-vector drive's ability to maintain tightly regulated speed and torque. If this signal is lost (i.e., if the encoder fails at the motor), then the drive may fault. Make sure that the encoder is securely and properly mounted to the motor. This encoder signal is compared to the commanded speed. By knowing what the motor slip is, a precise speed correction can be made in the form of new voltage output to the motor. This is made possible by utilizing high-throughput microprocessors to crunch the data and create a newly corrected output continuously. Thus, the flux vector drive is constantly correcting as the motor runs, much like an open-loop drive but on a higher level. This feedback now makes the flux-vector drive a true, closed-loop drive.

Many vector drives on the market are now being offered as sensorless-vector systems. Called *sensorless-vector control* (SVC), much of the control circuitry found in a closed-loop-vector design are present in the sensorless design, but without the feedback device. These drives do perform better than a standard volts-per-hertz model but must be furnished with pieces of data concerning the motor being controlled (no-load current, FLA, slip, motor reactance, etc.). An autotune with the actual motor is critical for a sensorless-vector-control (SVC) scheme. Importance is placed on the flux-vector drive's control algorithms for torque and speed control and also the feedback device. The last, but maybe most important, component of the flux-vector AC package is the induction motor. Attention must be given to the AC motor, which is being controlled by the flux-vector drive. Its nameplate and test data should be available so the correct parameters can be entered into the drive. Also, the motor, physically, will have to be equipped with a pulse generator and most likely an auxiliary blower for cooling at low operating speeds. It is usually better to procure the motor from the same supplier as the AC flux-vector drive, thus assuring compatibility. However, it is common for the motor needing vector control to exist in the field, with test data having long since disappeared. Sometimes even the nameplate has been painted over. Now the task of getting pertinent motor information into the vector drive is more of a challenge. This situation is not as bad as it seems. Actual current measurements to show no-load current (which is actually magnetizing current) and full-load currents are going to be accurate values. The motor has to be modified to accept a pulse generator anyway, and the motor's inductance and reactance values will have to be predicted. All this information now gets entered into the flux-vector drive's memory, and then the experimentation can begin. In short order, the drive and motor will have become matched. Running the motor/drive system over a wide range of speeds and loads will satisfy the complete range of operating parameters. Thus, even in these instances, a standard VFD can be converted or physically modified in the field to become a flux-vector drive. This usually can be avoided if the up-front application evaluation is done.

Flux-vector drives can provide better control over volts-per-hertz drives in the following areas:

1. *Speed regulation.* A flux-vector system can usually provide speed regulation to 0.01 percent.

2. *Torque at zero speed.* A flux-vector drive and motor can provide full, holding torque with the shaft at standstill. This means that full current is being provided to the motor and that auxiliary cooling should be supplied to the motor.

3. *Torque linearity.* Some applications require smooth torque throughout the speed range even during acceleration and deceleration. Torque pulsations can actually leave marks on material. The flux-vector drive exhibits smooth, linear torque regulations continuously, regardless of loading.

4. *Impact loading.* The flux-vector design responds better to load changes. Its response time, which is faster and has the encoder feedback, allows it to correct to these kinds of conditions much better than a volts-per-hertz control scheme.

Other considerations—some unfounded—concerning flux-vector-drive systems follow:

1. AC flux-vector-drive systems cost more than standard VFDs. This is true, so it must be confirmed that the application really justifies the extra cost. The major contributors to the cost increase are the motor with auxiliary blower, the added

control and feedback modules, and the pulse generator. These drives are not for
energy savings!

2. The encoder is an insignificant component in the system. Not so! It is vital to help
overcome the slip characteristics of the motor; in fact, it can shut down the entire
process. Affixing the feedback device concentrically with no misalignment and
with a good, flexible coupling is most critical. Special care in mounting these
devices is imperative.

3. Speed and torque requirements are needed. Determine beforehand the speed range
and the actual torque requirements, especially at the low speeds. Also, does the
load change frequently? Resolution of these issues will help select the proper size
and type controller to use.

4. What about a forward and reverse deadband? The AC flux-vector drive has no
deadband and can reverse direction quickly. This is done electronically and requires
no contact. The better question is, can the load be stopped quickly enough to
reverse?

5. How is troubleshooting the flux-vector drive different? The flux-vector drive is
built from the same blueprint as the standard, open-loop VFD. Therefore, many
issues and concerns are similar to that of the standard VFD. Troubleshooting of the
flux-vector drive is going to be very similar to that of the VFD but with some
control differences.

Field-Oriented Control

The *field-oriented controller* utilizes the base design of the volts-per-hertz VFD and the converter and inverter power bridges. It also is similar in design to the flux-vector controller in
that it wants to control the flux in the induction motor and also control the torque precisely. It can be a closed-loop design with encoder feedback, or there are also open-loop or
sensorless designs that can work very well. The field-oriented controller differs from the
flux-vector design in that it has the capability to control the flux and the torque independently from one another and thus gain precise control of the motor's torque.

The field-oriented-control scheme utilizes a high-bandwidth-current regulator (a physical component that makes up the control-board design) and corrects for torque error. This
error is derived from the encoder feedback and compared to the speed reference. From
here, a separate flux and torque reference is generated. The flux reference is compared to
actual-current feedback, and the torque reference is compared to values stored in the drive
from a previous autotune of the motor and drive. This autotune is usually done on commissioning the drive and motor and yields information such as motor-slip percentage, no-
load current, reactance, resistance, inductance, and other electric data. If an autotune is not
performed, then values for these parameters must be manually entered into the drive from
test reports or known values. The accuracy of this entered data will directly affect the drive
and motor's performance(s) when loaded.

As can be seen in Figure 3–31, the field-oriented controller counts digital pulses from
the encoder and uses that information as its speed feedback. If the motor speed is not where
it is supposed to be in that instant of time, the adaptive controller immediately corrects
(assuming that the load caused the speed deviation and more torque or current is required).
Likewise, the open-loop or sensorless version of the field-oriented controller can operate
much better than the standard volts-per-hertz design. However, not having the feedback
will mean that performance now is more dependent on that data entered into the drive

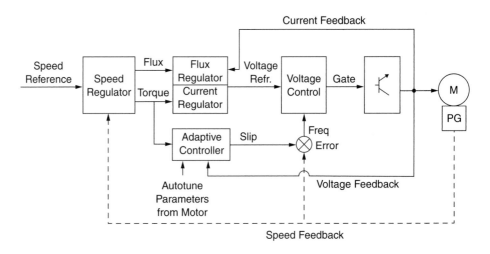

Figure 3–31 A field-oriented-drive circuit.

concerning the motor. Feedback will always help to make a drive-and-motor system operate at its optimum. Feedback devices have traditionally been an add-on component, which adds cost to the system. They have to be physically mounted somewhere on or near the motor, and a pulse-counting module must be provided within the drive. Wires also have to be routed from the pulse generator to the drive. All of this makes for more potential problem areas in the drive-and-motor system.

Direct-Torque Control (DTC)

The theory for direct-torque control (DTC) dates back to the early 1970s and combines the fundamentals of a process's ability to control itself and field-oriented control of an induction machine. Not until the past decade has this technology been expanded for the betterment of industrial-motor control. The technology utilizes the fastest-responding, digital-signal-processing hardware available and a model—complete mathematical understanding—of exactly how an induction motor functions. The net result is another high-performance drive-controller, very similar to field-oriented controls, which focus on the motor's magnetizing flux and the motor torque as two separate, controllable entities.

Direct-torque control works on the premise that *no* feedback devices are needed at the motor—no external excitation, no tachometer, no encoder, and no need for another physical piece of hardware that could fail. The DTC is actually very similar to a DC drive's dynamic performance in that it utilizes those actual motor parameters to produce the proper torque for the speeds desired. As is the case in the volts-per-hertz drive, the output voltage and output frequency are controlled and handled within the drive's gating and control algorithms, but such is not the case with the DTC device. The DTC is more concerned with what is happening inside of the induction motor and thus reacts according to those conditions. Those conditions are the motor torque and the motor's magnetizing flux.

With DTC, a tougher application that might require torque control at low speeds is more readily solved. Responses of 1 millisecond are typical for the torque response versus up to 25 milliseconds for DC drives and even some flux-vector drives. Many drive suppli-

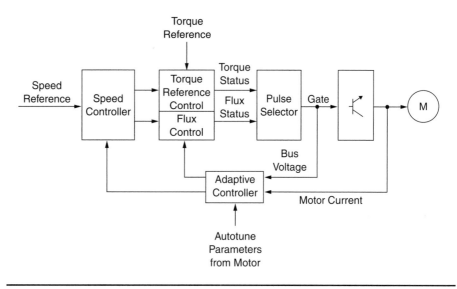

Figure 3–32 The direct-torque-control circuit.

ers of the DTC device claim full torque at zero speed, which is approaching the performance of servo systems, even with sensorless systems. The simplified control scheme is shown in Figure 3–32 for the DTC. Normal design calls for no feedback devices at the motor; however, if one were incorporated, the overall motor dynamic-speed regulation would increase to match servo performance.

CHAPTER 4

Installation Issues Concerning Variable-Frequency Drives

As has been discussed, solid-state VFDs are more than just speed controllers. Besides the actual drive hardware and software, there are many issues to address when installing them into any facility. The existing electric and mechanical systems need to be looked at to assure a successful VFD installation. Some critical installation issues relative to solid-state VFD systems are:

1. How will they be controlled and thus wired?
2. Complete circuit protection—breakers or fuses—must be incorporated.
3. Shaft voltages and overvoltage reflection are concerns.
4. Harmonic distortion and power quality are issues.
5. Have electromagnetic interference (EMI), radio-frequency interference (RFI), and grounding been considered for the VFD installation?
6. Where will the VFD(s) be physically located?
7. What about the heat the VFDs generate? How is it dissipated?

Many of these installation considerations interrelate to one another. Location of the drive plays a lot into minimizing maintenance and problems later. Proper routing of wires and power cable also must be carefully planned. Most importantly, the entire control and installation scheme should be completely thought out ahead of time.

Control and Power Wiring Methods

Variable-frequency drives are often located some distance from the motors they are controlling. The drive or drives may be located in an enclosure in a control room, in a motor-control center (MCC), in a basement (vault), or on a mezzanine. Therefore, it is common to locate an operator console at the motor or machine so that as adjustments are made, the operator can physically see the change at the motor. This operator console can be as simple as a start-and-stop pushbutton with a speed potentiometer or as complex as a full-color monitor with touch-screen capability. The latter is obviously much more elaborate and expensive and requires special communications schemes, local microprocessor ability at the

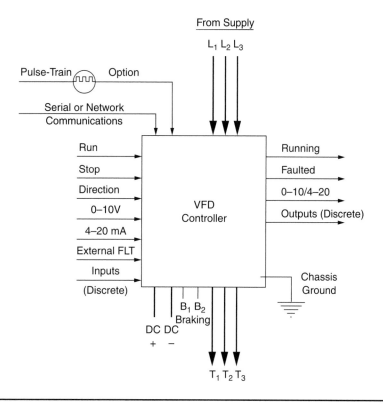

Figure 4–1 A sample VFD control scheme.

console, and possibly additional cooling at the console. These systems are typical in multi-story office buildings, which may have an energy management system to give a central command center, via a main console, the ability to view any fan or pump motor drive. Polling the drive can reveal drive status, speeds, currents, and so on. A sample VFD control scheme is shown in Figure 4–1. This wiring diagram is typical and covers commonly used control, power, and grounding schemes. Each drive manufacturer's manual should provide recommended wire-sizing and circuit-protection data. Tables 4–1 and 4–2 are included to aid in the sizing and voltage drops of aluminum and copper conductors.

A typical operator's console may house some basic VFD control devices. Those devices include push buttons for starting and stopping one or more drives, a maintained jog push button, and a drive reset push button. It is often convenient to have these push buttons lighted so they can be seen from a distance. Also on the console will be a device that can dial a speed into the drive. The speed potentiometer is very common and provides an analog signal proportional to requested motor speed. Every application's control needs are different and so too is its operator's station. All stations must have an emergency stop (E-stop) push button with a large, red mushroom head for safety. Other applications can utilize reverse switches, tension potentiometers, various lights as alarms, and so on.

This operator-console information often goes to a host computer before it is, in turn, sent to the drive. If other drives or other machines are affected by the operator command, the main computer may need to alert the other devices accordingly. Sometimes, the host

TABLE 4–1 Diameter, Resistance, and Weight of Aluminum Wire

Wire Size/ AWG	Diameter/ Nom Inch	Resistance in Ohms Per Thousand Feet	Weight	
			Lb Per M Feet	Feet Per Lb
1	.2893	.2005	77.02	12.98
2	.2576	.2529	61.08	16.37
3	.2294	.3189	48.44	20.65
4	.2043	.4021	38.40	26.04
5	.1819	.5072	30.47	32.82
6	.1620	.6395	24.15	41.41
7	.1443	0.8060	19.16	52.19
8	.1285	1.016	15.20	65.79
9	.1144	1.282	12.05	82.99
10	.1019	1.616	9.56	105
11	.0907	2.04	7.57	132
12	.0808	2.57	6.02	166
13	.0720	3.24	4.77	210
14	.0641	4.08	3.77	265
15	.0571	5.15	3.00	333
16	.0508	6.50	2.37	422
17	.0453	8.18	1.89	529
18	.0403	10.3	1.50	666
19	.0359	13.0	1.19	840
20	.0320	16.4	0.943	1060
21	.0285	20.7	.748	1340
22	.0253	26.2	.590	1690
23	.0226	32.9	.471	2120
24	.0201	41.5	.371	2700
25	.0179	52.4	.295	3390

Note: 1/0, 2/0, 3/0, 4/0, 250 MCM, 350 MCM, and 500 MCM sizes can vary based on the manufacturer's insulation thickness and whether or not there is armor or jacketing. VFD power cables in these sizes may vary in configuration, diameter, and weight.

controller monitors other conditions about the overall machine. It may calculate at what speeds and torques individual drives should run based on various inputs into it. Likewise, the main controller can download "recipe" information for various runs of product for that particular day. Sometimes the VFD is the slave in the system, and other times it is the master. In either case, the VFD is not the "dumb" device. Its microprocessor capability is being utilized somehow; to what extent is up to the user.

Harmonic Distortion/ Power Quality

Installation of solid-state VFDs requires the existing power supply to be employed. The facility's power supply and the quality of that power will dictate whether or not nuisance trips, electric noise, and harmonic distortion will become factors. Will the supply power be clean—free of other harmonic frequencies itself? Will there be frequent outages? Brownouts? Overvoltage or undervoltage conditions? Is the power supply a "stiff" source with high-interrupting capability? Installing a solid-state VFD may have to include instal-

TABLE 4–2 Diameter, Resistance, and Weight of Copper Wire

Wire Size/ AWG	Diameter/ Nom Inch	Resistance in Ohms Per Thousand Feet	Weight	
			Lb Per M Feet	Feet Per Lb
1	.2893	.1239	253.3	3.947
2	.2576	.1563	200.9	4.978
3	.2294	.1971	159.3	6.278
4	.2043	.2485	126.3	7.915
5	.1819	.3134	100.2	9.984
6	.1620	.3952	79.44	12.59
7	.1443	.4981	63.03	15.87
8	.1285	.6281	49.98	20.01
9	.1144	.7925	39.62	25.24
10	.1019	0.9988	31.43	31.82
11	.0907	1.26	24.9	40.2
12	.0808	1.59	19.8	50.6
13	.0720	2.00	15.7	63.7
14	.0641	2.52	12.4	80.4
15	.0571	3.18	9.87	101
16	.0508	4.02	7.81	128
17	.0453	5.05	6.21	161
18	.0403	6.39	4.92	203
19	.0359	8.05	3.90	256
20	.0320	10.1	3.10	323
21	.0285	12.8	2.46	407
22	.0253	16.2	1.94	516
23	.0226	20.3	1.55	647
24	.0201	25.7	1.22	818
25	.0179	32.4	0.970	1030

Note: 1/0, 2/0, 3/0, 4/0, 250 MCM, 350 MCM, and 500 MCM sizes can vary based on the manufacturer's insulation thickness and whether or not there is armor or jacketing. VFD power cables in these sizes may vary in configuration, diameter, and weight.

lation of extra hardware items such as isolation transformers, input reactors, or input filters into the overall circuit. Then there is the issue of the harmonic distortion that may be created by the drive itself, thus disturbing the power supply.

In any installation with electric supply, there are linear loads and, many times, nonlinear loads. A *linear load* can be defined as a predictable sinusoidal waveform generated by an electric load. Examples of this include a facility's incandescent lighting system and its other resistor and inductor loads. This is termed *predictable* because a relationship exists between the voltage and the current whereby the sine wave is *cleaner* and smoother. However, with the advent of rectifier circuits and power-conversion devices came *nonlinear loads.* Other nonlinear loads have also been introduced over time for one energy-saving reason or another. Lighting ballasts, which help to save energy in lighting systems, exhibit a certain hysteresis, which contributes to magnetic saturation thus causing the load to be nonlinear and the wave to be nonsinusoidal. Other nonlinear devices include copying equipment, electric-heating equipment, power-factor capacitors, resistance welders, and arc furnaces. Even computers, with their switching properties, are nonlinear loads. With these nonlinear loads come waveforms that are no longer nice, clean sine waves. The waves now carry additional frequencies and, actually, many portions of other waves that make the resulting wave distorted. The questions now become, what problems are caused by this condition and what can be done to correct the situation?

Since solid-state power converters (AC–DC) represent the largest contributor of harmonic distortion, there has been a significant rise in the awareness of harmonic distortion, especially as this equipment has proliferated. Variable-frequency drives, uninterruptible-power-supply (UPS) systems, and electric-heating equipment all convert AC to DC, or DC to AC, and by doing so create changes to the sinusoidal supply. This can cause definite interference problems with communication and computerized equipment prevalent in the industrial facility of today. A typical electric circuit in a factory today can have motors, drives, sensitive computers, lighting, and other peripheral equipment on it. Every factory is different, and every circuit within the factory is different. Thus, a complete analysis must be made of the system in place to properly identify the magnitude of the harmonic distortion and the corrective action required.

Each *harmonic* has a name, sequence, and frequency. The *sequence* is the phase rotation with respect to the fundamental in an induction motor. The *fundamental* is the 60-hertz AC sine wave. A positive-sequence harmonic generates a magnetic field that rotates in the same direction as the fundamental. A negative-sequence harmonic rotates in the reverse direction. The harmonic is a sine-wave-based component of a greater periodic wave having a frequency that is an integral multiple of the fundamental frequency. In simpler terms, we start with a pure, electric sine wave and evolve into a nonsinusoidal condition. The other harmonics are the integer multiples of the fundamental that "ride along" with the sine wave as unwanted frequencies. The second harmonic is a frequency of twice the fundamental, or 120 hertz. The third harmonic is three times 60 hertz, or 180 hertz, and so on. In a six-pulse system, the most critical harmonics to filter are the fifth and seventh, which can create unwanted heating conditions in the plant with regard to electric equipment. The fifth harmonic, a negative-sequence harmonic, can cause mechanical damage as well as electric damage if not filtered. It should be noted that even harmonics disappear when waves are symmetrical.

Table 4–3 shows the frequency and sequence of several harmonics, and Table 4–4 shows harmonic effects. The overall system that contains drive equipment exhibits a certain amount of total harmonic distortion (THD). This is due to the nonlinear loads that are produced by these static-power-conversion devices (AC and DC drives), arc furnace devices, some rotating machinery, and so on. These nonlinear loads create havoc with highly sensitive communications equipment and microprocessor-based equipment. If all loads were linear, then there would not be a need for the IEEE 519-1992 standard. Its existence means the disturbances associated with harmonic distortion are there and must be dealt with. A sample single-line diagram of a typical facility is shown in Figure 4–2. Here, linear and nonlinear loads are present. The single line should contain necessary data about the system in order to deal with the harmonics of that system.

In the early 1800s, Baron Jean Fourier, a mathematician, developed a series of formulas to work with nonsinusoidal waveforms. There must always be a fundamental component that is the first term in the sine and cosine series. This is the minimum frequency required to represent a particular waveform, typically 60 hertz. As we have seen, there are integer

TABLE 4–3   Classifying Harmonics

Harmonic Name	F	2d	3d	4th	5th	6th	7th	8th	9th
Frequency (Hz)	60	120	180	240	300	360	420	480	540
Sequence	+	–	0	+	–	0	+	–	0

Notes:

1. The fundamental is the 60-hertz frequency.
2. Sequence refers to phasor rotation with respect to the fundamental in an induction motor.
3. Zero (0) sequence harmonics (odd multiples of the 3d) are called *triplens.*
4. Even harmonics disappear when waves are symmetrical, typical for most electric circuits.

TABLE 4–4   Harmonic Effects

Sequence	Rotation	Effects (Skin Effect, Eddy Currents, Etc.)
Positive	Forward	Heating of Conductors, Circuit Breakers, Inductors
Negative	Reverse	Heating of Conductors, Circuit Breakers, Plus Motor
Zero	None	Heating of Conductors, Breakers, Motor, Plus Neutral

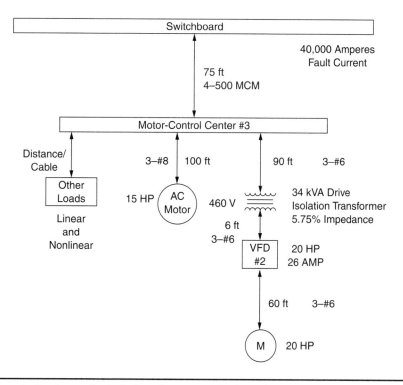

Figure 4–2   A sample single-line diagram of the linear and nonlinear loads in a facility.

multiples of the fundamental. In a six-pulse system, typical of a drive's converter, the real harmonics of concern are the odd-numbered harmonics not divisible by three. The most critical harmonics to filter in the six-pulse system are the fifth and the seventh. In a twelve-pulse system, the lowest producing harmonic is the eleventh. When static power converters convert AC power into DC power, there is a disturbance created both to the incoming supply power and to the output of the converter. This is due to the waveform being distorted from its original sinusoidal state. The effect of the AC is also felt in this power-conversion step. This distortion can have detrimental effects on other electronic devices depending on the severity of the distortion. Studies have given industry standards by which electricians can, at least, minimize or predict the harmonic content of a given system and take corrective measures. One such standard is the IEEE Standard 519. The 1981 version of the standard was referred to for many years and was finally revised in 1992 (another revision should be in process). In its present form, the standard has become more than a guideline; it now makes recommendations, meaning that the user with the harmonic problem now has some options. Prior to this, a facility had to live with problems or try to solve them on their own. Thus, harmonic-distortion "horror stories" and rumors about harmonics causing terrible electric catastrophes got started. It is important to note here that there is no one single solution to a harmonic-distortion problem, and it is not practical to think that just meeting an IEEE standard for allowable distortion levels of voltage and current will keep a facility clean. The issue of harmonic distortion definitely takes some study, and, usually, there is an associated cost for correction. To meet certain percentages of allowable distortion, more cost is typically incurred.

Total-demand distortion (TDD) is a value derived from the total rms (root-mean-square) harmonic-current distortion. It is typically expressed in a percentage of maximum demand load current. *Total harmonic distortion* (THD) is the full level of voltage and current distortion. It is sometimes called the *distortion factor* (DF) and is expressed as a percentage of the fundamental. It is the ratio of the rms of the harmonic content to the rms value of the fundamental. Thus, its mathematical formula is:

$$\frac{\textit{The sum of the squares of the amplitudes of all harmonics}}{\textit{The square of the amplitude of the fundamental}} \times \textit{100\%}$$

A shortcut method for determining or predicting harmonic distortion whenever PWM VFDs are installed is called the *short-circuit method*. By taking the available short-circuit capacity of the bus, where the drive will attach, and dividing by the drive's rating, a very good approximation of what amount of distortion to expect can be obtained. Figure 4–3 shows a graph on which—after calculating the ratio—a predicted amount of voltage distortion can be plotted with respect to common PWM and diode-based VFDs. The higher the ratio, the less the distortion. However, a full and complete harmonic analysis by qualified individuals is the most accurate *and recommended*. The harmonic analysis should neither be oversimplified nor taken so seriously that all effort goes into looking for every minute load contributing to the nonlinear-load total. The predominant loads will be apparent.

A definition of what is existing in the facility, knowing the facts about the electric power (can be found from the utility), and tabulating all the loads (linear and nonlinear) will not only help pinpoint existing problem areas but also help to predict problems when future equipment installations are concerned. Starting with a single-line diagram of all the loads and attributing values to each are the first steps. From here, a determination can be made

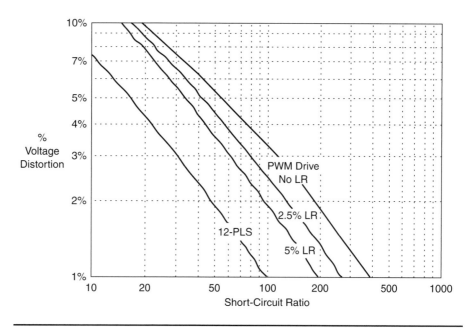

Figure 4–3   Graph of short-circuit-ratio method of estimating voltage distortion.

about what kind of filtering, if any, is required and a prediction given of what might happen if other loads are added to the system. A *tuned filter* is a specific combination of inductors, capacitors, and resistors so configured to provide a value of impedance to handle the one or more harmonic frequencies. The tuned filter has some costs associated with it and may not be effective if the plant's loads change dramatically down the road. It may be simpler and more cost effective to place a filter ahead of the only computer in the circuit rather than installing a more elaborate filter upstream. There is a relationship between the amount of harmonic filtering and the apparent costs involved. This has to be a consideration at some point, along with the ability to predict load changes on the particular system in the future as new pieces of equipment are introduced.

For any harmonic analysis, many entities play a role. The drive manufacturer, the electric utility, telephone company, and, obviously, the end user all have an interest. The telephone company is interested because of the telephone-influence factor (TIF). This is a complicated formula of weighted and nonweighted values regarding sine-wave components used to place a working value on telephone interference due to distortions to the current and voltage waveforms. The factory or plant, obviously, has a lot to say about this and any other sensitive pieces of equipment. The suppliers of the phase-controlled converters or rectifiers have much to offer also. It is important to include all parties along with someone who understands the issue of harmonics and can act as the consultant, or even the mediator, in solving problems and disputes.

To accomplish a study and harmonic analysis, the point of common coupling (PCC) must be selected. This is an important location in a facility's electric system, as it is the point where the harmonic-distortion measurement will be made. This point is usually at the secondary of a transformer where many parallel loads of the same electric system come together to connect to the main power supply. Regarding transformers, in general, a couple of comments should be made. First, many people are of the understanding that isolation transformers eliminate harmonics. This is not so, as harmonic currents pass through a transformer. Voltage distortion can be affected by the impedance of the transformer, but this can also be accomplished with a less-expensive line reactor, providing it has an equal impedance value. Secondly, transformers are designed to operate at 60 hertz. The harmonics tend to be present at higher frequencies, thus creating losses in the transformers in the form of heat. A transformer can overheat if subjected to currents containing high levels of harmonics. This has given rise to what is known as the K factor—or form-factor element—in transformer sizing in converter/inverter applications. Likewise, other electric equipment, and even the power cabling, is often derated due to harmonic content to the waveform. There can be a certain amount of heat caused by harmonic distortion. This value, while not dramatic, is still present. IEEE 519-1992 includes a cable-versus-harmonics derating chart for use on six-pulse systems. When designing an electric system where converting equipment will be used, attention should be given to this matter.

Not only will this power conversion affect the supply-power system, but it will also affect the output waveform's shape, most often to a motor. The occurrence is the presence of higher-frequency disturbances that now tag along with the useful voltage and current. These notches are called *commutation notches,* and their depth is of great significance. Also, the frequency with which these notches are seen is important. Basically, since controllers are changing sine-wave AC into pulses, this being accomplished by delayed conduction, harmonics are now produced. Solid-state devices are the culprits in these rectifier systems, with SCR and other thyristors being exceptionally notorious for high amounts of distortion because of their intermittent conduction. They appear as a short circuit to the system because they are not dissipating any energy when they conduct—good for efficiency but bad for harmonics and power factor.

IEEE 519-1992 sets limits on distortion limits for both voltage and current. Some are classified by the depth of the commutation notch, the area of that notch, and the THD in terms of voltage. Limits are also being set on how much the current waveform can be distorted as well. What all this means is that manufacturers of power-converting equipment will have to provide filtering where required; thus, system costs will go up. This is due in part because the harmonics are not necessarily harmful by themselves, except to other sensitive electronic equipment on the same electric system. Equipment gets hot; neutrals are not large enough, and they get hot; and power factor increases with harmonics. In summation, harmonics can, are, and will be blamed for a variety of problems in the plant, such as blown fuses, burned-up motors, hot or burned connectors, and hot neutrals. All point to a potential in-house harmonics concern. It is worthwhile to make an analysis and then take corrective action.

Overvoltage Reflection

Overvoltage reflection has the potential to damage many electric motors in a facility over time. The terminology goes by many names: reflected wave, standing wave, ringing, overvoltage reflection, voltage spikes, or the transmission-line effect. Many even classify this as *harmonics,* which is true in part with respect to distortion out to the motor. Whatever one might call it is not as important as what causes it and how it can be dealt with. Some up-front planning regarding the location of the VFD with respect to the motor may eliminate the need for most other solutions. One very important consideration when installing a VFD is its location. Cable distances from the drive to the motor can be potentially harmful to the equipment when fast-rise-time transistors are supplied in an inverter. Thus, locating the drive in close proximity to the motor that it is controlling is recommended. Though, this is not always possible, it is always much easier to route cable and wire than to reposition machines and motors. Additionally, many heating, ventilating, and air-conditioning (HVAC) retrofit applications start with an air-handling unit on the roof of a building. The first choice is to mount the drive up on the roof with the air handler and motor. This poses a problem due to the heat generated by the drive. Heat and VFD enclosures are discussed further in Chapter 5, but finding a drive-mounting place near the motor and under cover, or indoors, is recommended.

Reflected-wave or standing-wave conditions can result in premature motor failures. The breakdown of the motor's insulation is accelerated, and the motor life expectancy can be lessened. This insulation-breakdown phenomenon is described as repetitive, partial discharges of electricity seeking "the path of least resistance" within the motor. Eventually, a catastrophic condition results with a motor short circuit or corona, a condition where voltage ionizes in the air gap and arcs across the windings. Many times, smaller motors fail and are replaced without ever evaluating the reasons for the failure. These smaller motors—usually throw-away items under 10 horsepower—are at the highest risk.

Fortunately, there are various steps that can be taken to minimize the risks of these motor failures. Evaluate the particular installation, and take the appropriate action to protect the motor. New installations are easier to work with than are retrofit installations, but protective measures can be taken *to minimize the risks of premature motor failure:*

- Specify and purchase inverter-duty motors that meet NEMA specification MG-31.
- Keep cable lengths from drive to motor as short as possible.
- Install output reactors, filters, or parallel resistive-capacitive (R-C) circuits between the drive and motor.
- Lower the carrier frequency to reduce the quantity of pulses to the motor from the drive.
- Whenever new cable is pulled, consider using a VFD-rated cable to handle the extra heating and/or to remove the voltage spikes.

As can be seen in Figure 4–4, installing an output reactor between the drive and motor will slope off the waveform and actually lengthen the rise time of the pulses to the motor. A *reactor* is an inductor sized for the current rating of the motor/drive in the circuit. Inductors oppose changes in current are used often to limit rise times. In effect, the destructive forces will be reduced at the motor. This will allow for longer cable lengths but will also introduce a voltage drop that can cause a reduction in output torque. Consider this when installing the drive and motor. Installing an output filter between the drive and motor is another possible solution. An *output filter* is constructed as a combination inductance/capacitance "trap" to react to the overall impedance of the circuit it is attached to. Typically, a filter is sized and matched to a given circuit and will not work properly if changes are made to the circuit (for example, multiple motors that increase the load, larger capacity motors or drives, etc.). The output filter is often used for electromagnetic interference (EMI) and radio-frequency interference (RFI)—thus, can be additionally employed to protect the motor. However, its presence does amount to a voltage drop and net reduction of output torque to the motor. One other hardware solution is the installation of an R-C device at the motor. As is shown in Figure 4–5, this cost-effective device

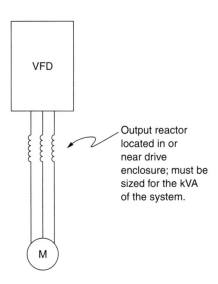

Figure 4–4 An output reactor can be placed in series with a VFD and motor to "slope off" the rise time of transistors.

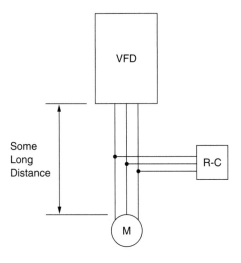

Figure 4–5 An R-C device can be placed in parallel with the VFD and motor to "trap" overvoltage spikes.

provides the resistive and capacitive match to the drive/cable run/motor combination to "trap" the voltage spikes before they hit the windings of the motor. There is usually no voltage drop and no reduction of torque, and the current waveform is maintained.

Shaft Voltages and Bearing Currents

Some premature motor failure has been pinpointed to bearing failure. Electric motor front and back end bearings can exhibit arc pitting and fluting damage, which eventually causes the motor to fail. Variable-frequency drives applied to motors have given rise to this par-

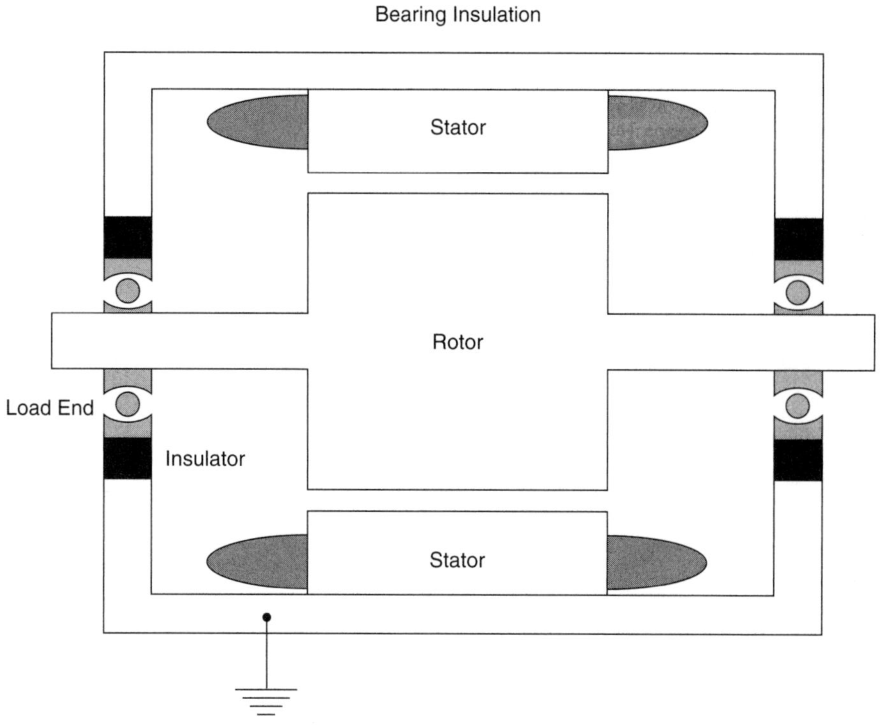

Bearing Insulation

Figure 4–6   Insulated or isolated bearings at the motor shaft.

ticular problem. Basically, the motor's bearings tend to act like a switch. Electrostatic voltages (5 to 30 volts) can accumulate on the motor shaft and cause the grease to break down. This then allows a path for the electric arc to take as potential builds between the shaft and ground. The arc damages the bearing, and the wear is exaggerated to the point of loose metal fragments becoming visible. The net result is motor failure.

While electric-motor shafts have been subject to electrostatic and magnetization effects for years, the application of solid-state AC VFDs to motors has brought more attention to this problem. Creating a balanced voltage with respect to ground is not possible with a solid-state AC VFD. The inverter's switching ultimately creates a potential within the motor, which has to discharge somewhere. This condition is sometimes called *common mode*. Common-mode voltages are *unwanted* voltages to ground that occur in various electric systems. Common mode is interference from a common voltage at the input terminals with relation to ground and is often expressed in decibels (dB). The interference can occur in various ways: as leakage currents in a poorly grounded system or as stray voltages in an unbalanced system seeking a ground point. Fast-rise times and switching frequencies of inverters contribute to this problem. However, there are solutions.

The simplest solution is to lower the carrier frequency of the solid-state AC VFD to less than 10 kilohertz or even lower (4 kilohertz). This solution is not going to cost to implement in terms of hardware and labor. It is most often a programmable feature of the drive. Other possible solutions are shown in Figure 4–6 through Figure 4–11.

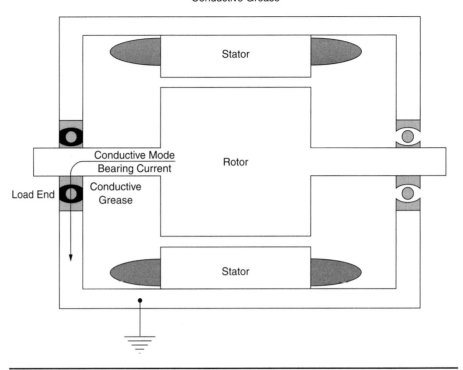

Conductive Grease

Figure 4–7   Conductive grease used to allow static voltages a path to ground.

All of the possible solutions to shaft voltages have pluses and minuses. Insulated bearings cost more than standard motor bearings, and there exists the possibility that the shaft voltages, while not transmitting via the bearing because of the isolation, could still build up. The shaft-grounding method, while effective, is not all that practical in the plant. Conductive grease does cost more than standard grease, and problems can occur if someone down-the-road puts standard grease on the bearings. The Faraday shield has a cost adder attributed to it and may affect VFD and motor performance. Another solution is to add an output filter between the drive and the motor, as was discussed earlier, to lengthen the rise time of the transistor pulses; yet, this still costs more in hardware and installation. Also, an isolation transformer—with a delta primary and wye secondary—will greatly reduce common-mode voltages within a drive-and-motor system. However, there is a cost adder attributed to its installation. Ceramic bearings could be an effective solution, but this is a newer concept and there needs to be more data collected over time to determine if this is truly viable. There are several solutions, and one not shown but worth trying is to lower the carrier frequency at the drive. To recap, *possible shaft voltage/bearing current solutions are:*

- Lower carrier frequency
- Insulated/isolated bearings
- Shaft grounding

Shaft Grounding

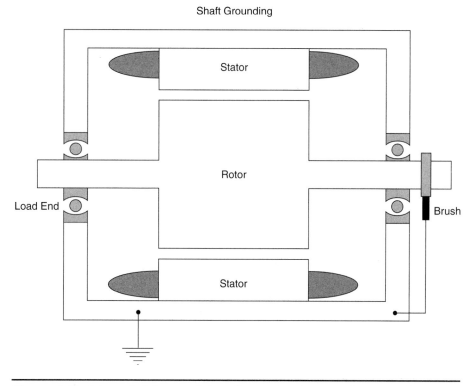

Figure 4–8 Motor shaft is grounded to allow static voltage a path to ground.

- Conductive grease
- Faraday shields
- Output filters
- Isolation transformer
- Ceramic bearings

Carrier Frequency

The fast switching of the VFD's transistors yields a quieter motor and smoother currents out to the motor. However, the side effects of this higher carrier frequency continue to "creep-up" in many installations. The more often a transistor is *on* and conducting, the greater the heating effect there will be, especially on the drive panel. Additionally, the fast-rise times needed to achieve the fast on and off states of these transistors lead to potential overshoot and overvoltage spikes. Since so many drive installations differ in terms of physical distance between drive and motor, conductor size, motor impedance, line capacitance, and so on, the effects of higher carrier frequency are often hard to predict. It is often worth the effort to raise or lower the carrier, or PWM, frequency value to see how it affects the drive-and-motor system. A scope may be necessary to view any changes. Also, when making these adjustments, make them in small, discrete steps, not in large steps. Above all, every owner of a VFD in a facility should *know* exactly to what value *every* VFD has its carrier-frequency parameter set to. It is best to know the starting point!

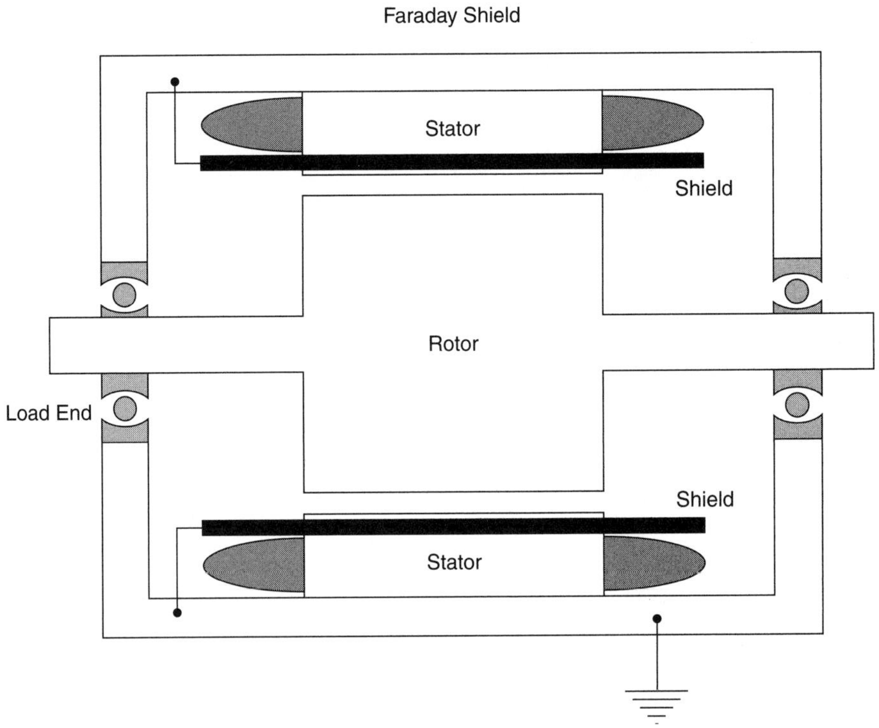

Faraday Shield

Figure 4–9 Faraday shield to pick up static voltages and route to ground.

Electric Noise, Electromagnetic Interference and Radio-Frequency Interference

With every electric circuit having some noise content in it, installing solid-state VFDs becomes very challenging. Oftentimes the noise cannot be traced to any one device or any one source. Many times, it is intermittent, which makes finding the problem even more of a chore. Electric noise, crosstalk, hash or trash, or even "garbage," as it is sometimes called, is the culprit in many electronic plant shutdowns. The key here is that years ago machines and processes had few components that were microprocessor based. The microprocessor and the low levels of voltage in the circuitry surrounding it are very susceptible to electric noise. With electronic data traveling bit by bit across a conductor at low levels of current, it does not take a whole lot of higher-level energy to disturb one bit or destroy it. This is basically what electric noise will do.

When the subject of electric noise is discussed, there are many ways to describe it, find it, and handle it. Describing electric noise is interesting. When discussing noise, we are mostly concerned with noise emissions and noise immunity, that is, where does it come from and how can the control system work adequately if noise is present? A *noise emission* is defined as electromagnetic energy emitted from a device. *Noise immunity* is the ability of a device to withstand electromagnetic disturbances. In looking at noise emissions, we find that most electric noise is manmade. There are natural emissions from the voltage coming off of all electric components; these are called *thermal noise emissions.* Thermal noise emissions rarely affect the electric system. Another type is the natural, atmospheric-type noise that is mainly attributed to lightning storms. The last type is the manmade noise; since it

Ceramic Technology

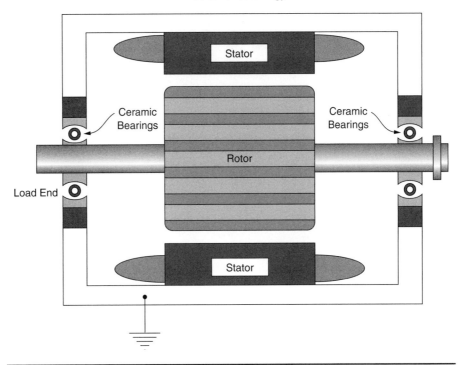

Figure 4–10 Ceramic technology has made strides into the motor-shaft-bearing realm.

is created by humans, it can be controlled by humans. This type is made up mainly of conductive, inductive, and radiated noise emissions.

There are basically five types of electric noise: conductive, capacitive, inductive, radiated, and that which has a shared impedance. All are shown graphically in Figure 4–12. *Conductive noise* is physically introduced into a circuit via wires or conducting materials. It is defined as either normal mode or common mode. Normal-mode electric noise is present between the conducting wires and the circuit reference. Common-mode electric noise is present between the circuit conductors and ground.

Capacitive or electrostatic noise is introduced into a circuit by a capacitive effect and is voltage based. *Capacitance* is the ability to store an electric charge when any two conductors are placed near one another separated by a dielectric. In essence, every wire or conductor placed near another will exhibit some degree of capacitance (the wire's insulator is the dielectric); as the voltage increases, so too will the capacitive effect. This capacitive, coupling effect has been around for decades and explains many of the problems faced in the plant with power wiring, overvoltage spikes, and noise. Capacitive noise can be minimized in VFD and motor installations by taking care in the routing of power cable. As can be seen in Figure 4–13, the input power cable should not enter the drive enclosure in the same conduit as the output power cables to the motor. Separate conduits—in and out—should be incorporated. After all, they have to separate somewhere outside of the drive anyway!

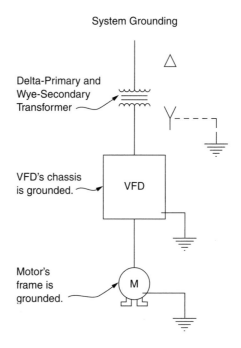

Figure 4–11 Proper system grounding will help eliminate static voltages and leakage currents.

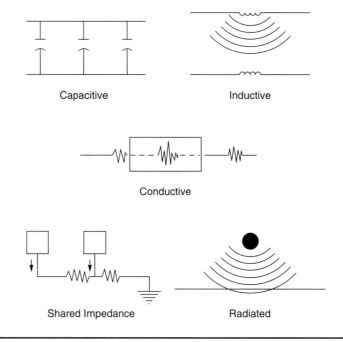

Figure 4–12 Electric noise types: capacitive, inductive, conductive, shared impedance, and radiated.

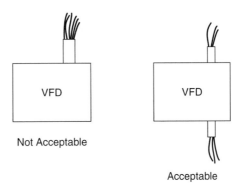

Figure 4–13   Route input power and motor cables in separate conduit.

Unfortunately, when dealing with VFDs and electric motors, the principle of induction is at work to make the whole process of motor and drive function. Without this principle, nothing will move! *Inductive or magnetic noise* is current based and is introduced into an electric circuit by induction. It is induced from coiled conductors and seeks other coiled conductors, thus creating yet another type of electric noise.

Radiated noise comes from that kind of emission that travels through the air. This RFI varies by distance from the source. Being able to pinpoint where the noise is originating from is obviously going to help eliminate the problem. Radio-frequency interference is discussed further later on with solutions provided, as it is a type of EMI, which is an issue with fast-switching, high-frequency, solid-state drives. *Shared impedance* is noise introduced into a circuit whenever two circuits share common wires. Ground loops are a major cause of this type of electric noise between multiple electronic devices.

There are many different types of noise-related phenomena. Being able to distinguish one from the other and taking the proper steps to avoid each is important. These types include harmonic distortion, EMI, ground loops and floating ground situations, power fluctuations, and RFI. Complete power outages and brownouts can also be thrown into the category of noise, because an interruption of service is an issue. If the interruption is short (a cycle or less), then maybe the sensitive equipment will keep running. If it is longer, then relays, most computerized equipment, and other electronic devices will shut down, thus demanding resets of the equipment throughout the plant. This can be costly. Each type of electric noise has its own cause, and each exhibits its own effect.

Radio-frequency interference (RFI) is actually an EMI at higher frequencies. It is an electromagnetic radiation where the source is farther away than the radiation's wavelength divided by two times pi. This is typically in the range of a few 100 kilohertz to one gigahertz. Radio-broadcasting frequencies are typically above 150 kilohertz and below 100 megahertz. Whenever current or voltage waveforms are nonsinusoidal, there is always the possibility of having RFI. Newer switching technology in converter and inverter devices has started to infringe on the high-frequency domain of the radio waves. By the laws of the Federal Communications Commission (FCC), it is forbidden to interfere with these signals. Switching frequencies, or carrier frequencies in drives, have reached values of 15 kilohertz with present-day drive technology. Since drives utilize the switching capability of different transistors, they have come under scrutiny. When current increases so quickly, as is

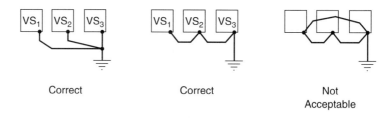

Figure 4–14 Avoid ground loops when installing multiple VFDs.

the case with today's power semiconductors, noise can be detected at a radio. Distance from the emitting device, shielding, and filtering are all factors relating to the existence of RFI. Other factors impacting RFI are the horsepower and current rating of the drive, the output and switch frequency, the impedance of the AC supply, and how well shielded the power modules are. Thus, the common methods used to reduce the effects of RFI are to (1) eliminate the radiating source, (2) shield either the source or the receiver, or (3) create some distance between the radiating source and the sensitive equipment.

Conductor lengths, common-mode chokes, differential inputs, and twisted-pair cabling, as well as selecting proper components will help to minimize RFI. Filtering by means of inductors and capacitors is presently a common way to suppress RFI emissions with respect to VFDs. This problem, in general, regarding RFI and drive equipment, is still ambiguous. Emissions and occurrences are rarely proven, as they are rarely considered. Enforcing the adherence to the laws and standards has not been paramount. As the occurrences increase, so, too, will the need to police the issue. One step is to clearly specify that the drive manufacturer must adhere to certain levels as set by the FCC. Then these levels must be measured after drive installation to see that they are met. By enforcing the issue, proper attention to the matter will be given.

Ground loops and poor grounding are often sources of electric noise. A ground circuit is supposed to route higher levels of current to ground rather than destroy valuable electronic components and shock humans. Unfortunately, this is also another route for low-level electric disturbances to travel. This ground route can send leakage currents and unwanted disturbances out to other sensitive devices, and it can also receive unwanted electric noise. A ground loop exists when there is potential, or EMF, between pieces of equipment whose grounds are connected in the same system. A typical ground-loop condition can be seen in Figure 4–14 and should be avoided. The best scenario is to ground each separate piece of equipment directly to its own ground. The trick is not to create a loop condition. This loop will allow unwanted signals from various sources to enter into a completely separate circuit, often a circuit whose equipment is very sensitive.

Grounding, when done solidly and properly, is a very good and necessary practice. However, grounds can become loose over time in environments where vibration and just simple activity around the connection occur. Also, grounds can deteriorate over time, especially in corrosive atmospheres. Checking and inspecting ground wires and all wires in a circuit are just good practice. Making sure the connections are constantly solid can save damage to devices that normally expect low levels of voltage and can also eliminate disturbances due to ground loops. Proper grounding of a transformer secondary is shown in Figure 4–15.

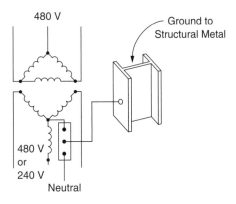

480 V

Ground to
Structural Metal

480 V
or
240 V

Neutral

Figure 4–15 Grounding of a transformer.

Shielding is the most common practice of protecting signal wire from electric noise. As discussed earlier in this book, there are several techniques that can be utilized when routing signal wire and the type of cable actually used. Coaxial cable provides an ample degree of shielding due to its composition. The outer casing protects the inner signal conductor. Coaxial cabling is a safe choice when electric noise is expected in a system, but not at too high of levels. Another common type of signal wiring is that of shielded wire. This should be used for any speed-reference signal into a VFD. Here, a three-conductor wire is encapsulated with a shield. This is tied to ground at one end only. This helps keep unwanted noise from getting through to the wires and prevents potential ground loops. Another method that helps minimize these intrusions of noise is the use of twisted-wire pairs. Shielded *and* twisted wire would provide an even better degree of protection.

A *shield* is any material placed between unwanted electromagnetic fields to minimize the transmission of these EMI fields. Another form of shielding is the actual installation of physical barriers within an enclosure housing multiple electronic components. Often times these barriers only have to be sheets of metal just separating compartments inside the enclosure. This should contain emissions, some natural and some manmade, from getting to the sensitive devices within the enclosure. It may be possible to locate the sensitive component or the noise-emitting component so that it can be located away from the other, external to the enclosure.

As solid-state VFDs have sensitive control circuits and these control circuits make or break the application, another means of protecting the integrity of these control circuits is by using optical, or opto-isolators. We have all heard the term *isolation,* but what is it and how is it incorporated? *Isolation* is, in a basic sense, completely separating two components in an electric system such that unwanted noise or signals can not pass between. One common means of achieving this is by optical isolation. Since certain diodes can emit infrared light, it has become an electronics industry practice to use that capability to isolate expensive electric components. By using the light emitter in conjunction with a light receptor, no physical connection has to be made. The light receptor is sometimes known as a *phototransistor.* In this way, logic commons in low-voltage-control circuitry are fully separated or isolated. This is a safe and effective means of enabling or disabling a circuit. However, it is recommended that when interfacing the logic of one manufacturer's product with another's that compatible isolation techniques are incorporated. This can avoid problems of incompatibility at the start-up of the equipment.

With light-emitting diodes (LEDs), it has become a practice in many low-voltage-level control systems to use optical isolation. Basically, as the electric current passes through the diode (LED), the light is emitted. A light receptor located adjacent to and dedicated just to that particular LED receives the light transmission and keeps the circuit flowing. The attractiveness of optical isolation is that there is built-in circuit protection. No actual electric current flows through this isolation point. Since control equipment is usually expensive and board-level components are not easily replaced (it is now customary to discard the entire board rather than troubleshoot it and replace one or two components), it is extremely important to protect the low-level-voltage components from all possible spikes and surges.

One of the most important factors involved is location, location, location! Placement of the physical drive with respect to the supply power, the ambient conditions, and the motor to control play critical roles in the success of the solid-state VFD application. Mounting the drive near the motor, in a cool environment, and supplying it with consistent and clean power will make the installation reliable and more than likely trouble-free. In addition, the life of the solid-state drive equipment, as well as the motor, will probably increase. Place the physical location of the VFD high on the list of installation priorities.

Location, Enclosures and Heat Dissipation

Variable-frequency-drive equipment needs a proper, safe, and sturdy mounting (start with *NEC®*, Article 110, Requirements for Electrical Installations). It is not acceptable to simply start mounting these electric devices, relays, starters, and other switchgear on machines and on the walls in the factory or office building. For one thing, this procedure is not safe. For another, it is unsightly and will probably lessen the life of the electric product. Many industrial users, especially those with not a lot of electric equipment in their plant, take the easy route and mount the VFD in the most convenient place in the building, which is usually right to the wall. They may even jam all the input and output cables into the same conduit into the VFD, which is *another shortcut not recommended;* input cable should go into its own conduit and output cable into its own, into and out of the VFD.

Many facilities have dedicated space for electric equipment. Control rooms that can be given restricted entry are common in facilities that have a substantial amount of electric product in use. Electric rooms, mechanical rooms and vaults, and motor control centers (MCCs), along with outdoor buildings, can all serve as ideal locations for drives and peripheral equipment. Depending on the size of the drive and the quantity, there may be a need to further enclose the drive or drives. In facilities where specific rooms or vaults are not available and wall space may be at a premium, a drive enclosure may be the only alternative. This would be some type of a box, floor or wall mountable and made of a metal material (preferably steel as it provides electric-disturbance protection) to withstand the local environment and provide a degree of protection to the workers and the product within. This has become the accepted solution for housing most electric product in today's factory. The VFD equipment does not necessarily have to be located right next to the motor it is controlling nor right next to the supply transformer (although keeping the cable distances short is recommended). Sometimes it is not possible to physically locate right next to a piece of equipment. This is what wire and cable are used for—to connect the two pieces electrically. Therefore, enclosures can be located away from the traffic and actual production areas in the plant, within reason. It still costs extra to run cable longer distances, but there are usually logical locations to place enclosures.

Once a suitable location is chosen, then it is important to select the appropriate enclosure type for the application and for the actual environment. To aid in this process, there exist standard ratings as established by NEMA for degrees of protection, both for indoor and outdoor enclosures. These are shown in Table 4–5. These ratings state the degree of protection required but do not always imply the recommended means of achieving that protection. For instance, an enclosure must be completely sealed and placed indoors. There may be heat-producing devices within the enclosure that will not work well (or at all) when the temperature increase within gets too severe. Thus, an air conditioner may be added to the enclosure for local cooling of the enclosure contents. Thus, the integrity of the sealed, indoor enclosure is not diminished. Likewise, the same rating enclosure could also tolerate simple muffin fans pulling air in, over the devices, and back out—this in lieu of an air conditioner. Costs, interpretation of standards, and common sense should prevail.

Since heat is produced by the VFD equipment, it must be "transferred" away, there are five basic methods of transferring that heat:

1. Surface radiation
2. Forced air
3. Heat exchangers
4. Air conditioners
5. Vortex coolers

A method should be selected for a given application's ambient conditions, its cost constraints, and its practical implementation.

Selecting an Enclosure

Factors to consider when selecting an enclosure for VFDs include not only the location but also variables that may require some control features. The following sections discuss some of those factors.

Total Heat Content   Drives produce heat. If multiple drives are to be placed on a panel and into an enclosure, then some calculations must be made in order to determine if the cabinet will be ventilated or cooled. Some drives can be configured such that their heat sinks can protrude out the back of an enclosure. This will virtually eliminate most cooling and ventilation requirements. Consult the drive manufacturer for the watts loss of the drive equipment. Figure 4–16 shows Fourier's equation for heat transfer, which is exactly how the heat loss for a residence is calculated. All practical examples use the common formulas:

$$°F = \left(\frac{9}{5} \times °C \right) + 32$$

$$°C = (°F - 32) \times \frac{5}{9}$$

$$\frac{BTU}{hr} = watts \times 3.414$$

Ambient Conditions   The temperature of the local surroundings of the drive has a direct effect on the overall heating of the enclosure. Avoid mezzanines, steam pipes, and hot

TABLE 4–5   **Environmental-Protection Classifications**

Type	Enclosure
1	Intended for use primarily to provide a degree of protection against contact with the enclosed equipment
2	Intended for indoor use primarily to provide a degree of protection against limited amounts of falling water and dirt
3	Intended for outdoor use primarily to provide a degree of protection against windblown dust, rain, sleet, and external ice formation
3R	Intended for outdoor use primarily to provide a degree of protection against falling rain, sleet, and external ice formation
3S	Intended for outdoor use primarily to provide a degree of protection against windblown dust, rain, and sleet and to provide for operation of external mechanisms when covered with ice
4	Intended for indoor or outdoor use primarily to provide a degree of protection against windblown dust and rain, splashing water, and hose-directed water
4X	Intended for indoor or outdoor use primarily to provide a degree of protection against corrosion, windblown dust and rain, splashing water, and hose-directed water
5	Intended for indoor use primarily to provide a degree of protection against dust and falling dirt
6	Intended for indoor or outdoor use primarily to provide a degree of protection against the entry of water during occasional temporary submersion at a limited depth
6P	Intended for indoor or outdoor use primarily to provide a degree of protection against the entry of water during prolonged submersion at a limited depth
7	Class 1, Group A, B, C, or D hazardous locations, air-break—indoor
8	Class 1, Group A, B, C, or D hazardous locations, oil-immersed—indoor
9	Class 11, Group E, F, and G hazardous locations, air-break—indoor
10	Bureau of Mines
11	Intended for indoor use primarily to provide a degree of protection against dust, falling dirt, and dripping noncorrosive liquids
12	Intended for indoor use primarily to provide a degree of protection against dust, falling dirt, and dripping noncorrosive liquids other than at knockouts
13	Intended for indoor use to provide a degree of protection against lint, dust, seepage, external condensation, and spraying of water, oil, and noncorrosive liquids

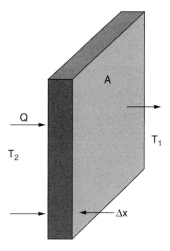

Fourier's Equation:

$$Q = \frac{k}{\Delta x} \, (A)(T_2 - T_1)$$

Where Q = Heat Flow
k = Thermal Conductivity
Δx = Thickness
A = Area
T = Temperature

Figure 4–16 Formula for calculating the heat loss of a VFD enclosure.

process areas. Variable-frequency drives typically work best in ambient temperatures of 0°C to 40°C.

Altitude Thinner air dissipates heat slower than does air at altitudes below 1000 meters (3300 feet). Consult the drive manufacturer to discuss drive deratings above specified levels.

Ventilation As is common to most drive enclosures, there will probably be slots and openings for heat discharge. Some cabinets will also incorporate forced-air ventilation with fans. Explore whether or not this is practical for the ambient conditions surrounding the enclosure. Dirt and dust being brought into the enclosure will collect on components and cause overheating. Likewise, if filters are used over any slots or openings, they will have to be cleaned routinely or similar overheating problems will occur.

Size Floor-standing or wall-mount units are available in different heights and depths. Doors should be able to be opened without obstruction.

Layout of Drives Drives within the enclosure should be located on the panel so that a maintenance person can gain quick and convenient access.

Carrier Frequency of the VFDs Lower carrier frequencies allow for less heat loss. Consult the VFD manufacturer manual for actual values at different levels of carrier frequency.

CHAPTER 5

Troubleshooting Variable-Frequency Drives

In many applications, the VFD is usually the first suspected component when a problem occurs. Until the problem is isolated, the drive remains the main suspect. To compound matters, not many in a facility can really pass judgment on the drive because they do not understand its makeup. Additionally, the manual may not be available, and a desperation call to the supplier of the VFD might have to be made, or the drive supplier's technician might have to visit (at some cost) in order to determine that the drive is truly innocent. Being the mysterious "black box" can have its downside.

Training on VFDs is becoming a business itself. There are so many drive manufacturers out there, and each drive offered is just a little bit different. This means that electricians *must* be trained on the equipment. They also have to understand the overall application and installation of the drive with respect to peripheral electric and mechanical equipment. The Occupational Safety and Health Administration (OSHA) standard, "Employee training requirements for the servicing and maintenance of electrical equipment," should be considered.

The VFD "system" is a mix of computerized, electric, and mechanical equipment. Thus, troubleshooting and getting to the root of problem(s) are sometimes difficult. A machine or process may not be operational because a parameter has been programmed incorrectly (computer-related operator error), because there is a short circuit in the motor and the drive's DC bus fuse is blown (electric problem), or because the conveyor belt is too heavily loaded and the drive has faulted due to an overload condition (mechanical problem). The bottom line is that the equipment must be brought back on-line immediately. This chapter reviews these facets of the drive installation and offers some methods for uncovering the problem and getting the machine or process back on-line.

Proper Setup

Before jumping to any conclusions, some facts need to be considered. First of all, has the equipment been set up properly when commissioned? Regarding the commissioning:

1. Was an authorized, trained electrician involved with the start-up of the drive?

2. Were the drive's setup parameters set mostly to factory default or were they changed to suit the application?

3. Was a list made of those parameters that were changed?

These questions can provide a certain who, what, when, where list of answers with which to start. Next, has the process or machine changed? Regarding the process/machine:

1. Is the process or machine (the motor) running at a higher speed?

2. Could the loads on the motor have changed?

3. Have ambient conditions changed around the process or machine?

The answers to these questions can provide clues as to what might be going on.

Fault Handling   Many times, the VFD will have some type of *fault* displayed on it? Ask questions first and seek basic information before tearing the "black box" (VFD) from the wall. Most often, the drive is doing what it is supposed to be doing—protecting itself and/or the motor from destruction. This may be a nuisance but it saves money for the user of the equipment. Finding the cause of the fault is necessary.

Why do VFDs and motors stop running? A motor can stop running or run in an unstable condition for various reasons (sound, polyphase-motor troubleshooting techniques should be employed here; Fluke Corporation provides good documentation for this). Is the motor overloaded? Is supply power present at the motor's input terminals—T_1, T_2, and T_3? Could the motor be short-circuited? Does the motor start but begin to run erratically or in an unstable manner? Remember that it is part of an electric circuit and it is doing work as part of a mechanical circuit. Isolate the problem. If the motor has become overloaded, reduce the load. A motor can only deliver so much output torque; it has maximum value. With the controller energized, check to see if supply power is present going into the motor. Many times a circuit breaker or contact has been left open upstream. Motors can fail and exhibit a short circuit. This can be due to bearing failure, overheating, or catastrophic winding failures due to corona, internal arcing, or shorts. A resistance check on the motor with external wires disconnected will help to isolate the problem.

As for the VFD, it is suspected even before the motor does or does not check out. Diagnosing the problem is more extensive, but some basic steps can be taken. First, if there is a fault displayed, then we have a good start. Most drives display the fault that takes them off-line, and many even retain that fault even if the display goes blank. Many times the fault is "tucked" into memory just before the drive goes off-line. Find the drive manual and keep it handy (inside the drive cabinet is recommended) and refer to it for fault diagnosis. Many drives will display a fault as a code, and this must be interpreted by cross-referencing it with the manual. Keeping a record of repeating faults is always a good idea, especially if a drive is ever returned for repair. Knowing the symptoms makes fixing the drive much easier. Additionally, many drives retain a history of faults within their memory.

If no fault is displayed or the drive will not power up, the troubleshooting is a bit more difficult. Difficult, yes, but checking for the presence of input power, the proper voltage rating, and control voltage is going to help us a lot. Likewise, checking for any output voltage to the motor may help to isolate the problem within the drive. Drives do have many more components than motors that have the potential to fail. Sometimes they do!

The aforementioned are merely starting points. Later on in this chapter, troubleshooting trees and diagnostic methods are provided to aid in keeping equipment on-line. Variable-frequency drives and motors can exhibit other problems that are harder to detect. This is going to be more apparent with the drive controller as it has more electronic circuits built in. Each drive manufacturer has an extensive list of faults inherit to their particular product. Good technical support from a manufacturer can be the best reason for selecting one drive over another. Not having to pay for it and getting that support 24 hours a day is even better. For those instances in which the technical troubleshooters are on their own, the following section deals with the more common faults and what typically causes them. Along with the fault and troubleshooting trees that follow is also going to be recommended maintenance information that will prevent some nuisance tripping and faults in the future.

Variable-Frequency-Drive Maintenance

Caution: All input power must be removed from any VFD before doing any maintenance. Also, dangerous voltages can exist in the VFD's capacitors. Check to see that all energy has dissipated through the capacitor's resistor circuit. If there is apparent damage to the physical drive parts, the circuit connecting the capacitors and the bleed-off resistors may have been destroyed. *Lethal voltages may still be present.* Lock out and tag out the input power source before doing any maintenance to the VFD.

A routine inspection should be done on an AC drive a minimum of every 2 to 3 months and more frequently if the environment around the drive is very dirty. The inspection and subsequent maintenance should include the following:

1. Visually check the VFD for any dust buildup, corroded components, or any loose connections.
2. Using a vacuum cleaner with a plastic nozzle to minimize damage to components while cleaning, vacuum dusty and dirty devices while brushing them off. Particular attention should be given to the rectifier-heat-sink areas. Excessive accumulation of dirt and dust will eventually lead to an adverse overheating condition (probably at a time of peak production). Cleaning the drive helps to avoid these types of problems.
3. Clean and retighten any loose electric connections.
   a. All fuse ends and bus-bar connections should be cleaned according to the manufacturer's instructions.
   b. In the case of screw pressure connectors, the power connections should be checked and retightened, if necessary. This may have to be done often during the first few weeks after installation.
   c. If any power modules are unfastened during the maintenance procedure, then the mating surfaces should be treated with the proper joint compound (per the manufacturer's recommended) before replacing. After replacing any power modules subject to certain torque values to their heat sinks, it is suggested that another check be made on the tightness of the screws and that they be retightened as necessary.
4. If the AC drive is mounted within an enclosure, check the fans and filters. If necessary, clean or replace the filters and replace the fan if the shaft does not spin freely.

Trouble-shooting and Repair of AC Drives

One VFD manufacturer's drive is physically constructed similar to another's. The hardware is virtually the same. Surface-mount-device (SMD) technology has helped make drive-circuit-board integrity very good, and many drive manufacturers are using this technique. Basic power conversion, available devices used, and standard wiring and protection techniques can also be assumed to be similar from manufacturer to manufacturer. However, control schemes, gate-firing circuits, packaging, and so on will definitely vary from drive to drive. Thus, it is mandatory that the manufacturer's installation, maintenance, and repair manuals be relied upon to perform any work on AC drives. Especially with digital drives and the numerous parameters that can be set and the extensive diagnostics provided, the manual will be the only place that will define many of the displays. This section discusses standard and common techniques along with practical applications and problem solving.

Caution: Electric shock can cause serious injury or even death. Remove all incoming power before attempting to do any work to any electric device! This is another reason to have the manufacturer's operation and maintenance manuals on hand, as they usually can safely take the repair person step by step through the situation. Whenever possible, work with another individual. Also, when using an oscilloscope, be aware that it, too, must be properly grounded. Always follow safe metering methods.

Spare Parts

Many times, in the "heat of the moment," when the drive has quit running and the factory is at a standstill, it is usually much faster to just replace the suspected bad component with a brand new component from the spare-parts inventory. Hopefully, there will be modules, boards, and plenty of fuses on hand. If not, then the importance of this issue needs to be re-evaluated. Spare parts can usually be shipped overnight from somewhere in the country, but nothing is better than to disconnect the bad device (work that has to be done even in a repair situation), go to the stockroom, pull the spare device, reinstall it, and get back up and running. Every manufacturer of AC drives will have a list of those parts for which spares are recommended to be kept on hand in the event of a drive failure. The best time to purchase these parts is at the time of drive purchase because this is the moment when the buyer has his or her best buying power. After that, the price of the spare parts becomes two to three times what they should have been. Besides not having them when they are needed most, the buyer ends up paying much more for them via overnight air freight.

For a typical AC drive, the following—as a minimum—should be on hand: various fuses at the ratings seen throughout the drive. Input fuses and branch fuses are a must as they protect the bulk of the drive's components from incoming power surges. These fuses do not have to be purchased through the drive supplier as long as the current ratings and sizes for the recommended fuses are matched. The power modules, transistors, or SCRs should be stocked as spares in groups of three, six, or twelve (for three-phase drives) and, obviously, in the exact ratings for the particular drive in place. Initially, these should be purchased through the manufacturer as there may be some characteristics that are inherent to the devices that are being used. The devices may have certain turn on/turn off values that may cause problems later if not matched accordingly. Often times, when an SCR or transistor fails, it is common practice to replace the other devices in the leg of the circuit because they may have become stressed and may be prone to quick failure.

Besides keeping a complete, spare VFD, other components that should be stocked as spares are the printed circuit boards for the drive. Unfortunately, these boards are not going to be obtained anywhere else but from the drive manufacturer. The control boards and gate-firing boards are the most common to stock, but consult with the drive manufacturer for any others. Also, consider electrostatic discharge (ESD) when handling any printed circuit boards. A grounding wrist strap is ideal but rarely practical when servicing a drive that is out of service. The next best procedure is to have an ESD-protected bag available, to minimize handling of the board(s), and to not touch small components on the board. Use one hand whenever possible to keep a path of voltage arcing from forming, and get the board into the bag as quickly as possible.

There may be some surge-suppression components called metal oxide varistors (MOVs) in place to minimize further component damage in the event of surges, and these can be stocked as spares but can be found in many electric hardware supply stores. They are the capacitors and resistors in the snubber network. Power supplies are other items that sometimes fail and are usually not manufactured by the drive supplier but rather purchased from elsewhere. Determine whether or not it is worth keeping these items in stock. Finally, to round out the necessary spare parts in the drive's inventory, any fans, feedback devices, special boards (for feedback, etc.), and any temperature sensors probably should be kept on hand if the drive is to be kept running 24 hours per day, 7 days per week.

Training

As another precautionary measure, continue to train various electricians in the plant on how to troubleshoot, replace, and repair the drive components. This training must be specific to the drive in place and can usually be performed by the drive supplier. Whenever the drive manufacturer can provide the field service, training, and additional required support, it is practical to get these services. However, they are usually very expensive and not everyone in the plant can get to be an expert on every drive. That is one reason why plants try to standardize on one or two particular drive manufacturers. However, this methodology is not always absolute because drive manufacturers change their designs over time and the user must relearn the product anyhow.

Diagnostics

The following sections cover many of the common problems, faults, and diagnostics that are inherent to most AC drives and their applications. Many drives have on-board diagnostics, which can help greatly in tracking the problem. Sometimes this is in the form of a display—either in alpha characters or by a numeric code number. The drive manual, which should be stored in a pocket on the drive door and not in someone's office, should be consulted.

When a drive is down, the first thing to do is to kill incoming power and lock it out (review OSHA standard for lock-out/tag-out practices). Be wary of drives that contain capacitors, which can carry charges for a period of time. Many drives carry a light-emitting diode (LED), which indicates when it is on that the capacitor charge has not been fully bled off. *Test these capacitors using a voltmeter to ensure that they are fully discharged.* Upon opening the drive cover or door, the initial checks should be to look for any apparent signs of internal component damage. That smell of fresh electrically produced smoke is a good sign. Next look for any burned wiring or components. Some components may even swell, so look for these—especially the capacitors. Any loose or disconnected wires should be noted, as well as any marks on the drive-enclosure walls from apparent electric arcs. Once the visual inspection is done, then begin further troubleshooting and tracing from the suspect point.

Power-Bridge Diode and Transistor
Testing

Converter Bridge:
Diode Module Resistance:
Disconnect input and output power wiring from VFD. Using a voltmeter or digital multimeter (DMM), place (+) leads on L_1, L_2, or L_3. The (–) lead should be placed on the VFD (+) DC bus terminal. Normal resistance will be 10–100 ohms (consult VFD manufacturer's manual). A bad diode module would be indicated by a 0 or infinite reading. If using the diode drop scale, then 0.3 to 0.7 is to be expected for a good module.

Inverter Bridge:
Transistor/Diode Module Resistance:
Disconnect input and output power wiring from VFD. Using voltmeter or digital multimeter (DMM), place (+) leads on T_1, T_2, or T_3. The (–) lead should be placed on the VFD (–) DC bus terminal. Normal resistance will be infinite (consult VFD manufacturer's manual). A bad transistor module would be indicated by a 0 reading. If using the diode drop scale, then 0.3 to 0.7 is to be expected for a good module.

Figure 5–1   *Converter and inverter bridge testing.*

Check all input fuses to see if any have blown. This may be visually apparent, or a physical check with an ohmmeter may be required. Either way, it must be determined whether the problem lies before, at, or after the fuse section. Many times, the fuses are blown and there is still a second or third problem with the drive. This is where the challenge of troubleshooting begins. Again, some drives may have the capability of saving multiple trips and faults. Use this VFD diagnostic tool.

Since most AC drives contain diodes and transistors, it is common to suspect these devices whenever there appears to be a problem. Many drives have the capability to display, when faulted, which transistor has failed to turn on or has been forced off abnormally by excess current. One probable cause might be that a high-inertia load has been accelerated too quickly. In this case, merely setting a longer acceleration time will solve the problem. Another cause could be the voltage boost. While a nice tool for hard-to-start loads, it is also a sensitive tool. Reducing its value will overcome motor losses and keep the transistor from failing. The last probable cause—and certainly not the easiest to correct—is the failed transistor. To determine if a transistor module is faulty, the resistance across its terminals must be measured with an ohmmeter (see Figure 5–1). Measuring normally with the one-times range, readings can be obtained that, when compared to the manufacturer's acceptable and nonacceptable levels, will indicate whether the transistor needs replacing. Transistors commonly come in modular form and are fairly simple to remove and then replace. The same procedures should be followed for any of the converter-bridge diodes.

Variable-frequency drives or today reveal much more than the drives of yesterday. When the drive trips, there is usually a self-diagnosis from the unit itself. Some go so far as to provide the actual text on a terminal screen, which tells the troubleshooter where to troubleshoot. Many drives trip—or fault—for the same reason, but the solution may mean looking at a certain terminal or adjusting a certain parameter for that particular drive. Therefore, the specific drive manual must be referred to for this kind of information. The common drive trips—sometimes even called alarms or errors—follow.

Troubleshooting Trees

Motor Will Not Run: When the motor will not run (Figure 5–2 and Figure 5–3, troubleshooting tree #1):

- *The causes could be:* no line supply power; no control power; output voltage from AC drive too low; stop command present; no run or enable command; no reference; some other permissive not allowing drive to run; or motor faulty drive.

- *Troubleshoot and isolate:* Check circuit breakers and contacts on input power side. Check to see if stop command is absent when trying to run motor. Check for run, enable, or other permissives being present. Try another AC drive if one in place does not respond.

Overvoltage: When there is an overvoltage fault (Figure 5–4, troubleshooting tree #2):

- *The cause could be:* DC bus voltage has reached maximum trip level.

- *Troubleshoot:* There is too short of deceleration time for load inertia; supply voltage is too high; motor may be overhauled by load.

Overcurrent: When there is an overcurrent fault (Figure 5–5, troubleshooting tree #3):

- *The cause could be:* Output current to motor has instantaneously reached too high of a level.

- *Troubleshoot:* Increase acceleration time, check transistors and motor

Main Fuse(s) Has (Have) Blown: When the main fuses have blown (Figure 5–6, troubleshooting tree #4):

- *The causes could be:* Bad transistor(s) or diode(s); short or bad connection in power circuitry.

- *Troubleshoot:* Check devices and replace if found defective; trace and check connections in power circuitry.

Overtemperature: When a VFD overtemperature occurs (Figure 5–7, troubleshooting tree #5):

- *The causes could be:* Heat-sink temperature sensor has caused trip; cooling fan has failed; drive ambient temperature has risen to extreme levels.

- *Troubleshoot:* Check thermistor with ohmmeter; if bad, then replace. Are heat sinks dusty and dirty? They should be clean. If drive is in an enclosure, are there filters? Are they clean? If not, clean or replace them. Is there a cooling fan? Is it operational? If not, then repair it.

Troubleshooting Tree #1

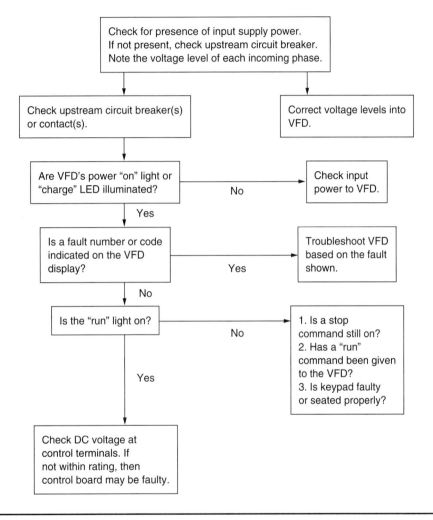

Figure 5–2 Troubleshooting tree #1.

Undervoltage, Low Line Input, or Power Dip: If undervoltage, lowline input, or power dip occurs (Figure 5–8, troubleshooting tree #6):

- *The causes could be:* Input fuse(s) blown; low input voltage; power outage.
- *Troubleshoot:* Check and replace fuse if blown; check to see that input voltage is within drive controller's allowable high and low range; check for possible problems with utility; sometimes lightening can cause this fault.

Troubleshooting Tree #1 (Cont)

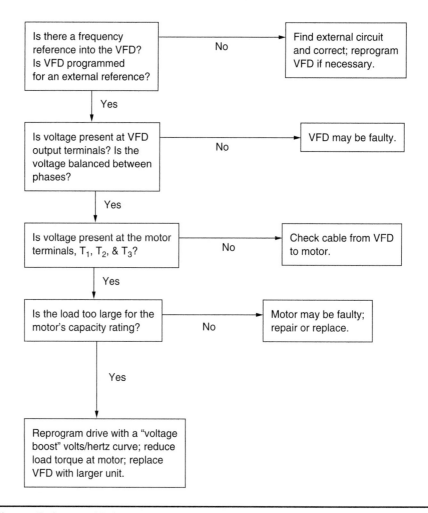

Figure 5–3 Troubleshooting tree #1 (continued).

Overload or Sustained Overload: If an overload fault or a sustained overload occurs (Figure 5–9, troubleshooting tree #7):

- *The causes could be:* Incorrect overload setting; motor cannot turn or is actually overloaded. Is motor hot?

- *Troubleshoot:* Check overload parameters and change to fit motor loading. Load may be too large for motor; a larger motor may be required or the gearing may have to be changed. There is a jam in the drivetrain; clear it.

Troubleshooting Tree #2

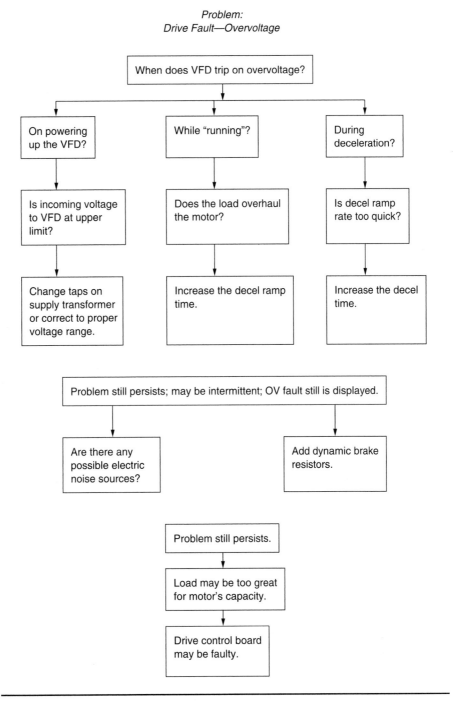

Figure 5–4   Troubleshooting tree #2.

Troubleshooting Tree #3

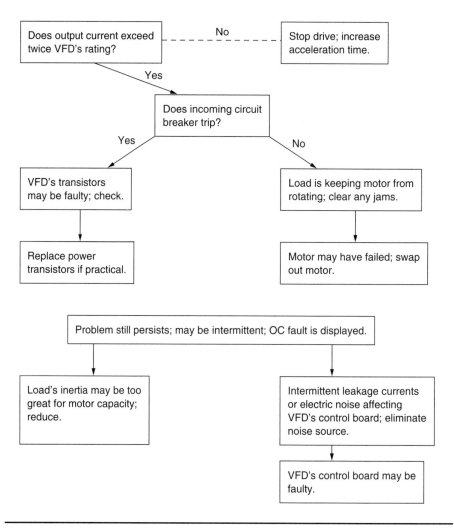

Figure 5–5 Troubleshooting tree #3.

Troubleshooting Tree #4

Problem:
Fuse Fault or Blown Fuse(s)

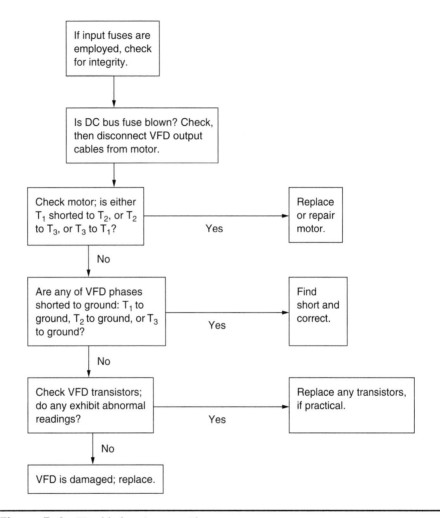

Figure 5–6 Troubleshooting tree #4.

Troubleshooting Tree #5

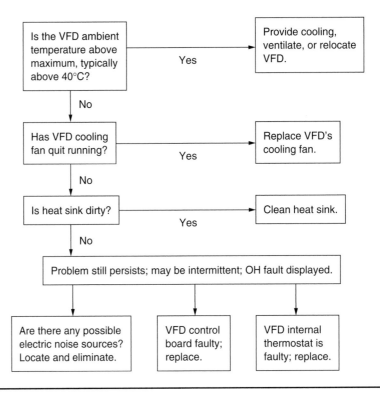

Problem:
Drive Fault—Overheat or Overtemperature

Figure 5–7 Troubleshooting tree #5.

Central Processing Unit (CPU) or Microprocessor and Communications Error: If CPU and communications errors occur (Figure 5–10, troubleshooting tree #8):

- *The causes could be:* Loose connector(s), EEPROM not seated in socket, bad control cable or wiring harness.
- *Troubleshoot:* Reseat suspect components (boards, electrically erasable programmable read-only memory (EEPROM)s, connectors, etc.); reset VFD to factory defaults.

External-Drive Fault: If an external-drive fault occurs (Figure 5–11, troubleshooting tree #9):

- *The cause could be:* Some condition outside of the physical VFD is causing the VFD's circuits to fault and prohibit running the motor.
- *Troubleshoot:* Check external devices; correct.

Troubleshooting Tree #6

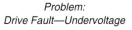

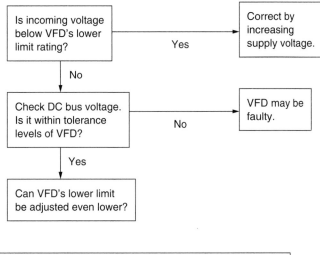

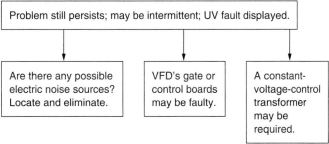

Figure 5–8 Troubleshooting tree #6.

Other Troubleshooting Options

If the speed at the motor is not correct or the speed is fluctuating:

- *The causes could be:* Speed reference is not correct; speed reference may be carrying interference; speed scaling is set up incorrect.
- *Troubleshoot:* Ensure that speed-reference signal is correct, isolated, and free of any noise; filter noise on line and/or eliminate noise; recalibrate speed-scaling circuit.

If the peripheral relays or control communication circuits are tripping:

- *The causes could be:* Drive carrier frequency is too high; other electric noise is present.
- *Troubleshoot:* Lower carrier frequency; eliminate electric noise.

Troubleshooting Tree #7

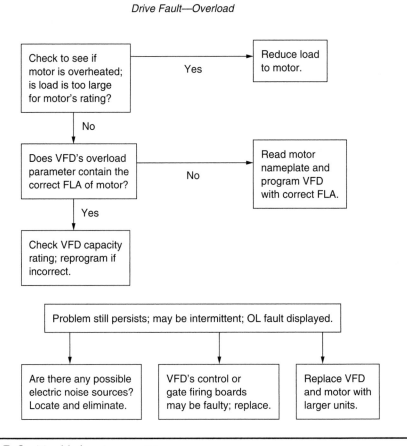

Problem:
Drive Fault—Overload

Figure 5–9   Troubleshooting tree #7.

If the motor stalls or a transistor trip occurs:

- *The causes could be:* Acceleration time may be too short; a high-inertia load or a special motor may be in use.
- *Troubleshoot:* Lengthen acceleration time. Readjust volts-per-hertz pattern for application.

If there is a short-circuit or earth-leakage fault:

- *The causes could be:* Motor may have a short circuit; power supply may have failed; excess moisture in inverter or motor.
- *Troubleshoot:* Check motor and cables for short. Have motor repaired; replace power supply.

Troubleshooting Tree #8

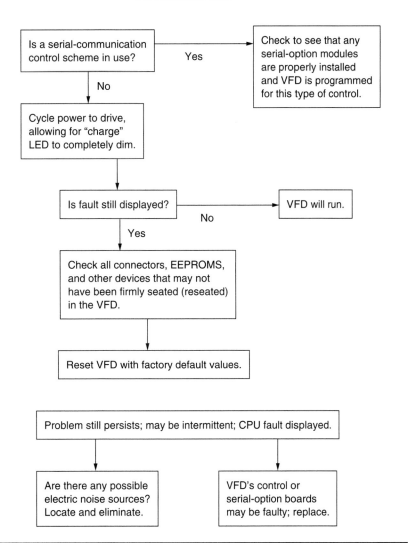

Problem:
Drive Fault—CPU or
Communication Error

Is a serial-communication control scheme in use?

Yes → Check to see that any serial-option modules are properly installed and VFD is programmed for this type of control.

No

Cycle power to drive, allowing for "charge" LED to completely dim.

Is fault still displayed?

No → VFD will run.

Yes

Check all connectors, EEPROMS, and other devices that may not have been firmly seated (reseated) in the VFD.

Reset VFD with factory default values.

Problem still persists; may be intermittent; CPU fault displayed.

Are there any possible electric noise sources? Locate and eliminate.

VFD's control or serial-option boards may be faulty; replace.

Figure 5–10 Troubleshooting tree #8.

Troubleshooting Tree #9

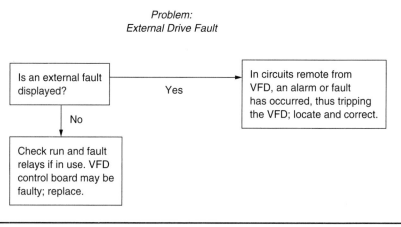

Problem:
External Drive Fault

Figure 5–11 Troubleshooting tree #9.

There are many other faults that can be exhibited by any given VFD-and-motor installation. Digital AC drives can always contribute the ever deadly CPU failure or show no response at all when the main microprocessor board is not functional. Cycling power to the VFD is worth trying whenever these "unexplainable" lockups occur. Sometimes this is also called a *watchdog-timer fault* with the main control board. With digital drives, if the drive can automatically tune itself with the motor and an operational error is detected, then the autotune sequence is aborted and an alarm is made. Additionally, any drive controllers that are set up to accept other external feedback, and it is not present, may trip upon powering up. As for motors, they can overspeed or run away, but usually the drive controller can be set up to trip at a maximum speed for protection.

Measurements and Tools for Troubleshooting Variable-Frequency Drives

Caution: Make safe measurements. Never exceed the working voltage, even if the measuring equipment has the range for it (especially with scopes). As mentioned earlier, troubleshooting drives and motors can be extensive. Starting with the simpler, easier areas can help speed up the process. Checking for loose connections, load and application changes, and programming errors within the drive can save time. Also, the accuracy of the measurement, whenever troubleshooting, will greatly aid in the diagnosis. With VFDs, *floating measurements*—those not grounded—are preferred; this will minimize the introduction of noise from grounded instruments. As will become evident, also, electric noise does become a factor in VFD operation and nuisance tripping. Using shielded meters, clamps, and probes is also highly recommended.

In Figure 5–12, a sample PWM-equivalent circuit is provided along with sample waveforms. As discussed earlier in the text, the input diodes provide a constant-voltage DC bus from which the transistors switch. This value is typically 1.414 times the input voltage. The capacitor filter network is there to help maintain a constant, ripple-free DC voltage for the inverter section.

The most common tool used by the electrician when troubleshooting the VFD is the meter. Which type to use is important. Analog meters give an output-voltage reading very close to what the PWM drive displays and close to what the predicted volts-per-hertz curve

Pulse-Width-Modulated Inverter

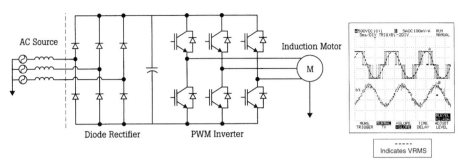

Theory of Operation

- DC converter section supplies constant DC level.

- Root-mean-square motor voltage is varied by the width of the PWM pulse.

- Motor-drive-signal frequency is controlled by the modulation frequency.

Figure 5–12 Sample PWM-equivalent circuit with waveforms (Reproduced with permission from Fluke Corporation).

is to be. However, analog meters do not have the International Electrotechnical Commission (IEC) 1010 safety rating and are not used too much anymore. Digital multimeters and rms meters read high. They read high because they are responding to all the frequency components of the modulated pulse. An analog meter has a low-pass filter, which allows for a much closer and accurate reading. The digital meters are not giving a *false* reading but rather its best reading based on the entire frequency spectrum and how often it samples. An average meter reads the output voltage of a drive high but not as high as the rms meter. It gets closer to the analog meter and what typically is displayed. The average meter reads the entire spectrum but does an average calculation. A scope or scopemeter can make the measurement more accurately because a low-pass-filter probe can be used and the volts-per-hertz value can be obtained with the AC rms volt setting. Additionally, the output-voltage waveform can be inspected for any overvoltage conditions. Figure 5–13 illustrates the waveforms expected from the output of a drive with a standard probe and a low pass-filter.

Checking the output voltage can be done with any of the aforementioned devices as long as the person making the measurement understands the reading he or she is getting and is consistent. If the reading is high, the percentage that it reads is recorded as high and then all other readings should be treated accordingly. Another key measurement to make is that of the three channels for voltage-imbalance conditions. As is shown in Figure 5–14, a voltage imbalance will typically create a current imbalance and VFDs will often fault under these conditions. If there is no voltage imbalance but there still remains a current imbalance, then this may signal that there is a problem with one of the motor's windings.

Adjustable-Speed Drive

Troubleshooting

- Motor Voltage Measurements (cont)
    - Use low-pass-filter probe with ScopeMeter to provide rms voltage readings of the PWM drive signal.
    - ScopeMeter can also be used to make V/Hz measurement.
    - Use 41B Power meter to provide rms voltage and frequency measurements of the PWM drive signal.

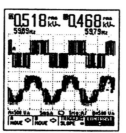

Top waveform made with standard probe.

Bottom waveform made with low-pass-filter probe.

Figure 5–13 Standard waveforms expected from the output of a VFD, measured with and without a low-pass filter (Reproduced with permission from Fluke Corporation).

Troubleshooting

- Motor Voltage Measurements, cont
    - Check for voltage imbalance (< 3 %) first. Then
    - Check for current imbalance (< 10%).

$$\% \ (V \ or \ I) \ Imbalance = \frac{Max. \ Deviation \ (V \ or \ I)}{Average \ (V \ or \ I)} \times 100$$

For Example:

	449				
	470				
	+462	(2)	$\frac{1381}{3} = 460$	(3)	$\frac{11}{460} \times 10 = 2.39\%$
(1)	1381				

Figure 5–14 Voltage and current imbalances (Reproduced with permission from Fluke Corporation).

Overvoltage reflections at the motor terminals have become a major concern within the drives industry. With transistorized inverter technology, faster rise times for the devices to improve efficiencies have also created a potential problem. The potential for overvoltage spikes is increased. Additionally, as these overvoltage conditions are reflected at the motor terminals, long cable lengths can aggravate the problem. Figure 5–15 reflects this concern and shows allowable cable lengths based on the increase over normal peak voltage. Many drive manufacturers make complete disclaimers on allowable cable lengths and do not allow for flexibility.

Troubleshooting

- Overvoltage Reflections at the Motor
  Terminals
  - Caused by excessive cable length and
    fast rise times of the drive signal.
  - Industry trend is for faster rise time
    drives to improve efficiency.

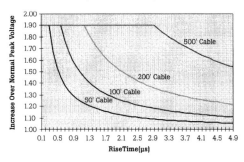

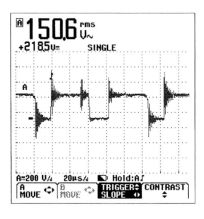

Figure 5–15 Overvoltage reflection at motor (Reproduced with permission from Fluke Corporation).

Making measurements with a digital multimeter or analog meter does not indicate the severity of the overvoltage problem. A scope or scopemeter can reveal the reflected voltages caused by fast rise times and long cable lengths. Checking for this condition at installation will help predict and pinpoint possible premature failures. Figure 5–16 shows a normal PWM waveform and a PWM waveform with reflected voltages. Overvoltage conditions can damage the first few turns of the motor windings and cause motor failure due to corona. If a motor is to be repaired, it should be rewound with insulated wire that can withstand substantial temperature rise.

However, all is not lost with respect to overvoltage situations. The electrician should look to see if the cable run can be shortened or make sure it is routed the shortest way possible between the motor and drive. If a new motor is used, make sure it is inverter duty and meets the NEMA MG-31 specification, which states that sustained voltage spikes of 1600 volts and rise times of 0.1 microseconds are tolerable. Lowering the carrier frequency of the drive is another possible solution, but there may be trade-offs. Audible noise, additional motor heating, and loss of usable torque can occur. Yet another solution is to use filtering or reactors in the circuit. Add-on (more hardware) solutions to incorporate into the circuit include the low-pass filter, series reactor, or R-C impedance matching filter at the motor terminals. These filter and reactor solutions are shown in Figure 5–17.

Placing an output reactor or filter between the drive and motor has the effect of snubbing or subduing the ringing and the spikes. However, the down side is that the reactor and filter act as voltage drops and a loss of torque at the motor is possible. This would only be a factor if the drive-and-motor sizing was marginal for the load(s). Another consideration in utilizing filters or reactors is where to physically place them. They have to be located in a safe, suitable place. Lastly, there is always the issue of cost. All these hardware solutions cost some amount, and this must be taken into account. Lowering the drive's carrier frequency is always worth trying first. It may have a positive effect and it is basically free.

Troubleshooting

• Overvoltage Reflections at the Motor Terminals

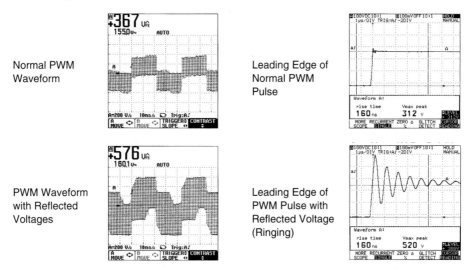

Normal PWM Waveform

Leading Edge of Normal PWM Pulse

PWM Waveform with Reflected Voltages

Leading Edge of PWM Pulse with Reflected Voltage (Ringing)

Figure 5–16   Normal PWM pulses (top), overvoltage in pulses (bottom). (Reproduced with permission from Fluke Corporation.)

Pulse-Width-Modulated Inverter

Troubleshooting

• Overvoltage Reflections at the Motor Terminals
  – Filtering Remedies

Filter Remedy 1: Typical effect of a low pass filter at the inverter output as measured at the motor terminals

Filter Remedy 2: Typical effect of the series reactor measured at the motor terminals

Filter Remedy 3: Typical effect of an R-C impedance matching filter measured at the motor terminals

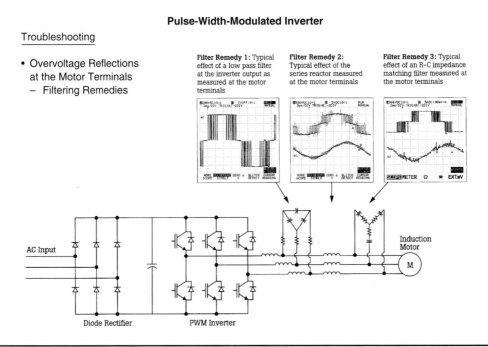

Figure 5–17   Remedies for overvoltage reflection (Reproduced with permission from Fluke Corporation).

During the motor-manufacturing process, winding machinery can create nicks when applying the insulation to the motor windings. Any holes, even microscopic, will allow a path for electric currents and thus eventual failure over time. Now it is necessary to specify a more substantial motor, the inverter-duty model, which meets MG-31 specifications (discussed earlier in the book). Additionally, lowering carrier frequencies has a positive effect on the motor. Insulated-gate-bipolar transistors (IGBTs) are the drive technology standard at present. The switching frequencies—or carrier frequencies—are not moving into faster realms as of yet. More emphasis is being placed on efficiencies of the transistors and thus the reduction of heat sink sizes.

Tips on Troubleshooting Variable-Frequency Drives

The VFD is protecting both itself and the motor that it is controlling. When it shuts down, it is not trying to put a company out of business; rather, it has detected a problem and that problem needs to be corrected. Sometimes, the drive trips can be a nuisance, especially if they occur often and other computerized equipment stays on-line in the plant while the drives go down. Some devices are more sensitive than others, but filters, snubbers, and capacitor networks can be utilized to correct for nuisance tripping.

Troubleshooting a drive system can be extensive and is not always straightforward. The operative word is *system,* because troubleshooting the drive may be only half the battle. For example, it may appear that there is an overload fault at the drive but in actuality there is a problem at the motor, or even in the machine, that must be found. A hands-on electrician with good mechanical and electric aptitude is sometimes needed, even though the drive is furnishing some data about the problem. All in all, drive troubleshooting is many times application and machine troubleshooting. Because the drive is the least-understood component, it is the easiest and thus the first component blamed. So often, a drive manufacturer's service person is called to the jobsite only to find that the problem was in miswiring, or motor short, or any number of nondrive-equipment failures. The bill for the service trip still has to be paid. A good, trained electrician or operator could do some productive troubleshooting and diagnosis beforehand. This could possibly save costly downtime and service bills.

Having the proper documentation, keeping good records, and having proper tools make troubleshooting VFDs much easier. At least two sets of manuals should be resident at a facility. One set should go in or near the physical drive on the factory floor. The other copy should be kept in an office or company library. Next, if drive parameters have to be set or programmed, then a copy of these should also be kept in both places. These settings or values may have to be re-entered if there is a dramatic problem with the drive. Also, regarding the issue of entering the drive's program, this should be a limited-access or by-password-only situation. Too many people having the ability to make changes will make for possible confusion. It is best if only one or two individuals can get into the drive to make changes. As for recordkeeping, a log—if not a standard feature of the drive hardware—should be kept to track when or if a drive faults and what was done to correct the fault. In this way, trends can be determined and lost production time can be fully tracked. Also, this information will be very useful to the drive manufacturer and/or service person.

Concerning the proper tools to have when troubleshooting the drive, this can get expensive. Tools also should include spare parts, and, again, nobody wants to purchase and stock expensive items unless they absolutely have to. This decision is mainly based on the quantity of drives and the value of lost production time. The expertise of the in-house electricians is another factor. Most facilities end up with one or two electricians who become the resident experts on the VFD. Many drives come today with a means of getting in and interrogating or making electronic changes. This may be a standard device, or it may be an optional handheld device. It is worth having around. Besides the standard current probes, volt-ohmmeters, and occasional pot tweeker (yes, some drives out there have many potentiometers that may need adjusting), a digital oscilloscope capable of storing images is always useful. Checking waveforms can help pinpoint problems both into and out of the drive. Also, sometimes it may be necessary to put a strip-chart recorder on a drive circuit to monitor line conditions over a period of time. This can again help to pinpoint trouble areas. Another tool that is recommended is a digital tachometer or handheld "tach"; it comes in very handy to measure the speed of a given motor. Since, these VFDs are asked to control the speed of the motor, it is sometimes necessary to physically see if the desired speed is being delivered. Also, the digital, handheld tachometer can be useful in determining speeds of other machine-rotating components and other motors. This data can be used to select motors and gear reduction for a given system.

Becoming intimately familiar with the type of drive and knowing the value of certain critical programmed parameters, its control and fault scheme, and the machine or application are critical components of any successful drive installation. This is true both for the user and for the supplier of the drive equipment. Complete training on the product is necessary. And, as previously mentioned, good documentation and tools can go a long way to keeping the drive(s) running. It should also be said that many drive installations, once started up, rarely if at all fault or cause problems. They are applied and installed with some up-front attention. As time goes on, the VFD is gaining some overdue respect.

The application of VFDs is where many engineering and technical disciplines meet. Electric equipment cannot always make up for shortcomings in the mechanics of a system or machine, and vice versa. The troubleshooter has to be aware of this. Electronic products are constantly being asked to do more and perform more functions. This is possible with digital products, but it also can make the products more complicated and harder to understand. This is not a problem as long as the electrician stays current with the technology and keeps learning. The opportunities are many. There are well over a hundred suppliers of VFD products in the world today. Each one's product is similar to the next; yet, there are also many differences.

The VFD's operation and service manual is as important as the drive hardware itself. Too often, the VFD runs for months or years without incident and then one day there is a problem with the drive. At that moment, invariably, no one can find a copy of the operation or maintenance manual. *It should be always left in or at the VFD enclosure.* Without the manual, the plant or factory is in jeopardy of having the drive stay out of service until whatever is ailing it can be fixed. The reason the manual is so important is that electronic, digital drives have become so powerful that they actually contain too much information. One recommendation is to write down the fault codes and tape them to the drive-cabinet door for quick reference. Another is to make a list of the parameters changed or necessary for operation. This way, getting back on-line can be expedited.

Gathering data concerning the drive is relatively easy, and the electrician should not be intimidated by the drive's complexity to gather it. Access to many circuits can be gained by the drive to analyze and monitor. With the aid of the microprocessor and extra memory, drives can be programmed to not only monitor many faults but also keep a log or history of them, when they occurred (the electronics allow for an on-board clock for elapsed-timer use), what they were, how and if they were reset, and so on. This data can be saved as the drive runs, and, upon a drive fault, the most recent data is retained in battery backed memory or to EEPROM. It can be retrieved by an electrician even with the drive down. This kind of diagnostic and troubleshooting tool is invaluable not only to the user's personnel but also to the manufacturer of the drive's service people.

Obtaining good information can mean getting the VFD equipment on-line much sooner. It is often necessary to telephone the drive manufacturer and talk to its technical support people. In the appendix of this handbook is a chart to keep track of drive faults and incidents. It is a place to log and track drive faults (many facilities actually tape these to the drive cabinet door). Having as much data as possible concerning the drive, drive type, serial number, size, and software modification or revision level is vitally important—just as important as the symptoms and fault information from the VFD. Knowing the VFD and where its critical parameters are set is imperative.

CHAPTER

6

Metering of Variable-Frequency Drives

Troubleshooting variable-frequency drives can be accomplished many different ways. This is because there are multiple elements at work in the VFD's circuit: electricity running through the VFD and motor, a software-based microprocessor somewhere, several external control functions, and a load (mechanical in nature) that wants to resist movement altogether. No wonder troubleshooting drives has become a challenge. The tools available to the expert VFD troubleshooter should include a scope, a digital multimeter (DMM), the VFD's own monitor device for fault and diagnostic data, and up-to-date documentation.

For starters, a VFD normally provides an indication of what has happened and why it has faulted. These faults happen for various reasons and usually can be explained. Refer to this book's Chapter 5 on fault handling and troubleshooting. But troubleshooters want to make physical measurements. Unfortunately, if that troubleshooter does not know what is going on inside the VFD and what is coming out of it, then the readings observed may seem erroneous and meaningless. Most experienced electric motor electricians are prepared to deal with traditional three-phase motor failures. Place a VFD in the circuit, and things are much different.

First of all, *safe measurements* are the rule. Know the voltage category that you are dealing with. All test equipment has a safety rating. Work on deenergized circuits whenever possible, and apply proper lock-out and tag-out procedures. Always use protective safety glasses and insulated boots. Try not to hold the meter in your hands (set it firmly in a position easily viewed). Use the three-point test method:

1. Using the proper meter, test a known live circuit.

2. Then test the circuit in question.

3. Finally, test the known, live circuit again to make sure the meter is still OK.

Always hook the ground lead first, and then connect the hot lead. This procedure should be reversed whenever removing the leads. Proper methods are important with VFD troubleshooting, also. Refer to OSHA Regulations (Standards - 29 CFR), "Working on Exposed Energized Parts" when in doubt of a procedure.

Input Waveforms

The input to the variable-frequency drive is sinusoidal. This input can be checked with an analog, DMM, true rms, or average meter. It is best to know what this voltage is coming into the VFD, if present, and if all three phases are balanced. Once inside the VFD, the three-phase rms power changes. Some voltage is stepped down to a lower control voltage, 24 volts DC, 15 volts DC, 10 volts DC, or even lower. Know those terminals on the drive that can be accessed for the presence of these voltages (this is where proper documentation comes into play). The voltage that is not stepped down is rectified through the diodes, and a DC bus voltage should be present. The DC+ and DC− terminals will typically read approximately 325 volts DC on a 230 volts AC supplied drive and 650 volts DC on a 460 volts AC supplied drive (these values occur when power is applied but no output is engaged). This waveform, when viewed, is straight DC, possibly with some rippling effect from the AC input.

DC Bus

The DC bus is a critical section of the VFD. It is the true link between the converter and inverter sections of the drive. Any ripple must be smoothed out before any transistor switches "on." If not, this distortion will show up in the output to the motor. The DC bus voltage and current can be viewed through the bus terminals. Other monitor functions are going on within the bus. Ground-fault conditions, overvoltage, and overcurrent conditions are constantly monitored. If the drive's control circuit is functioning well, then the trouble-shooter usually does not have to look at these specific test points. The precharge circuit is also part of the DC bus network. Here, a set of capacitors in conjunction with a start contact and some internal resistors provide a constant voltage for the inverter transistors. This precharge circuit is shown along with some other DC bus components in Figure 6–1.

Output Waveforms

The voltage applied to the motor terminals from the VFD's output, however, is nonsinusoidal. As is seen in Figure 6–2, the voltage wavefrom is made up of many pulses at a fairly consistent amplitude. This can make for some very interesting readings from a true rms meter, which has become fairly standard for most troubleshooting electricians. Additionally, the voltage measurements taken by other meters, analog and average, yield yet another different reading, while some meters do not react or provide a reading at all.

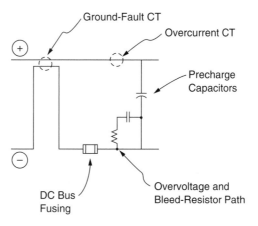

Figure 6–1 The VFD's DC bus.

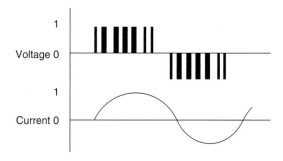

Figure 6–2 Pulse-width-modulated voltage and current waveforms.

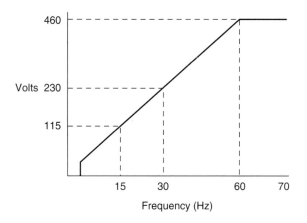

Figure 6–3 The measured output voltage of a VFD should follow a constant volts-per-hertz pattern.

The explanation is that in power-related voltage and current waveforms, there exist harmonic contents of the fundamental, 60 hertz wave. True rms meters tend to measure some of these unnecessary contents, which yield higher readings than the average, or mean, meter. The average value of each PWM pulse corresponds to the instantaneous value of the sine wave that the meter expects, thus confirming the higher reading from the rms meter. Most digital multimeters are rms type and respond to the high-frequency component of the drive's output waveform and thus produce the higher readings. The good news is that these readings are typically consistent with the meter through the entire speed range. Thus, as is shown in Figure 6–3, if a VFD is operating within its constant volts-per-hertz range, then full voltage should be at full speed, half voltage should be at half speed, and so on. Many times a meter can be applied to the VFD's output at full speed to get an indication of the meter's reaction and net reading. Note this indication and follow by changing drive speeds and watch the voltage. Compare this to the VFD's displayed output voltage.

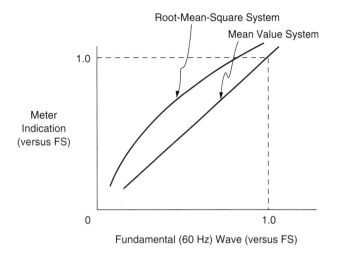

Figure 6–4 Voltage measurement of a VFD—comparing rms readings to mean—or average—readings.

Many troubleshooters prefer using an average or analog meter to measure VFD output. This is because the coil in the meter responds in the same fashion as does the motor—to the low-frequency component, not the high. Even a true rms meter can be used if it is equipped with "an average sensing/calibration" setting. This acts as a low-pass filter and will yield more expected results. Again, check the meter at full speed, at full output voltage, and through the speed range to see how it responds.

As was mentioned earlier, there is good news. The troubleshooter does not have to go out and purchase special accessories or metering equipment just for VFD metering. Understanding the output readings of the equipment at hand will suffice. Typically, the basic reasons for making voltage measurements at the VFD output terminals (or at the motor terminals) is to (1) check for the presence of any voltage at all, (2) check to see if that voltage is too high or too low, and (3) check to see that all three legs are balanced (an indication of a short at the motor or phase imbalance from inverter). If the voltage is not present, then that explains why the motor is not rotating. If voltage is too low, then there may be bad connections or serious voltage drops. If voltage is too high, then there may be overvoltage spikes present that need to be addressed.

In comparing the average and rms meters, there can be a form factor attributed to each. The graph in Figure 6–4 shows the indication versus the fundamental wave relationship. This yields the form factor as a ratio of rms to average. If it is a nondistorted sine wave, then the form factor is 1.0. If the value is off from 1.0, then that indicates a degree of wave distortion. Therefore, the true rms meter will give higher readings than the average meter. Knowing the starting point and what to expect will keep speculation to a minimum.

Relative to output current, the waveform should look fairly sinusoidal with a PWM drive's output. This wave will vary depending on loading and unloading of the motor, the actual carrier-frequency setting of the VFD, and the type of transistors used in the VFD's inverter section. Expected current waveforms should look like that shown in Figure 6–2.

Converter Bridge

Diode Module Resistance

Disconnect input and output power wiring from VFD. Using a volt-ohmmeter or a digital multimeter (DMM), place (+) leads on L_1, L_2, or L_3. The (–) lead should be placed on the VFD (+) DC bus terminal. Normal resistance will be 10–100 ohms (consult VFD manufacturer's manual). A bad diode module would be indicated by a 0 or infinite reading. If using the diode drop scale, then 0.3 to 0.7 is to be expected for a good module.

Inverter Bridge

Transistor/Diode Module Resistance

Disconnect input and output power wiring from VFD. Using a volt-ohmmeter or a digital multimeter (DMM), place (+) leads on T_1, T_2, or T_3. The (–) lead should be placed on the VFD (–) DC bus terminal. Normal resistance will be infinite (consult VFD manufacturer's manual). A bad transistor module would be indicated by a 0 reading. If using the diode drop scale, then 0.3 to 0.7 is to be expected for a good module.

Figure 6–5 Power-bridge diode and transistor testing.

Power-Module Testing

As previously seen, the converter section of the VFD is primarily made up of diodes, which allow the passage of voltage in one direction only. The inverter section is made up primarily of modules that are each made up of a transistor and diode in combination with each other. These configurations are shown in Figure 6–5. Thus, a physical check can be made of each power module in the VFD to verify the integrity of each device. To make the following power-module tests, the VFD must be disconnected from the supply power and the motor must also be disconnected from the VFD's output terminals. In doing so, erroneous readings will be eliminated due to the resistance from the external components.

The power modules can be metered relative to their resistance values or to the presence of a functioning diode and with the diode drop scale on a meter. To perform an accurate resistance test with a volt-ohmmeter, the power-module expected normal values must be known. This can be provided by the VFD manufacturer. If these are not known, then any readings showing a zero or infinite can be classified as abnormal readings indicating a short or open condition. Again, consult with the VFD manufacturer for proper, normal values. For input-module testing, the actual metering is accomplished by placing one lead on the bus terminal and then placing the other lead on each of the L_1, L_2, and L_3 input terminals, one at a time. The same procedure is duplicated for the output terminals, T_1, T_2, and T_3, one at a time. Testing in this manner, allows checking through the DC bus and through each diode module for resistance across it.

Another method of testing the power modules is by using the diode drop scale. As is shown in Figure 6–5, the diodes have a path that the voltage follows through the module and into the DC bus. Each input terminal and each output terminal of the VFD is checked in the same way that was done in the resistance check. Diode-drop-scale values should read between 0.3 and 0.7 typically, and they should be consistent. The measurement should be done twice at each device, changing the polarity. In this way, a preliminary diagnosis of the

VFD's critical internal components can be done and a decision can be made regarding whether a new drive should be installed or whether the transistor and diode modules should be replaced.

Drive types and drive sizes carry with them power modules that have different resistances and values. It is important to have this information on hand from the manufacturer to properly assess the functionality of the VFD's converter and inverter sections.

Harmonic Analysis

Often times, when considering the installation of a device that can convert AC to DC such as a VFD, an area of the plant where the VFD is to be located should be looked over. Develop a complete single-line diagram and analyze all the equipment on the same circuit. Calculate the total linear and nonlinear loads. Look at the impedances of transformers and reactors in the circuit. Additionally, obtain the short-circuit-current data from the utility. Predicted harmonic distortion is easier to correct for than the distortion that arises after the equipment is installed.

If it seems as though the entire plant could be affected, then all inductive and capacitive elements will have to be factored. All of these elements will have to be attributed with some value of impedance. Any power-factor-correction systems must be accounted for, because typically a power-factor capacitor—when affected by resonant oscillating currents—can actually make the power factor worse in the system. Thus, it should be made known what value of power factor must be held. Also, if the capacitors are switched, it must be made known when the switching occurs.

If symptoms of harmonics seem to exist, then it is best to first ascertain whether or not any rectifying equipment is in the plant. Then, in a step-by-step process, it is necessary to list and document all the electric components on a particular circuit. This *plant map* can be invaluable in totaling loads and impedance values; plus, it will indicate where sensitive equipment is in the plant and allow for planned filtering, if required. Likewise, the utility should be able to provide data relative to the source of power (from them) and the available short-circuit current, along with known impedances for upstream circuitry. After all, the power companies are partially driving the requirements of harmonic filtering and power-factor correction.

When testing for harmonic distortion levels, certain test equipment is required and the method of testing must be determined. Needed will be a current transducer, voltage leads, and some type of device to display and analyze the waveform. An oscilloscope, a spectrum analyzer, a harmonic analyzer, or a distortion analyzer will be required. This device will allow the user to take measurements of electric wiring to determine harmonic distortion levels. This device can display, from the readings, waveform shapes or graphs of a select harmonic. All this data can then be downloaded to a computer for further compilation. The digital analysis can be performed using either the fast Fourier transform technique or by using a digital filter. Both methods allow for the collection of spectrum data, which can be analyzed. This data can be compared to allowable limits and acceptable values.

Checking Pulse Generators

If a VFD is equipped to accept encoder feedback or pulses from another feedback device such as a magnetic pick-up, then it may be necessary to check for the presence and count of these pulses from that motor-mounted pulse generator. The VFD will not operate properly if these pulses—on and off states from a pulse generator external to the VFD—are not correct. To perform this test, the input power must be removed from the input terminals to make the proper connections to the lower-voltage, more sensitive, pulse-counter module.

Using a scope, connect the common to the test point labeled *ground*. Connect one channel to *PA* and the second channel to *PB*. Select a screen with a grid of 2 volts/division (amplitude) and 50 milliseconds/division for frequency. Now, input power can be reapplied, but do not command the VFD to run. With the motor unloaded, the motor shaft can be turned by hand. There should be pulses on both channels. They should read at a 5-volt level, separated by a half pulse, or 90 electric degrees, and the duty cycle should be 50 percent. This should provide a clear indication as to whether or not the pulses are coming from the encoder.

CHAPTER 7

Sizing and Selection of Variable-Frequency Drives

Applying solid-state VFDs involves addressing several issues. As has been shown, the location of the drive, the way it will be controlled, and the sizing are some of the key issues to consider. Asking and answering many questions about the application are appropriate. Full and complete investigation of the application will help understand the loads, control parameters, and general operation of the motor with the drive. Over the years, there have been all types of *application forms* devised by engineers and drive manufacturers that attempt to list (or ask) the relevant questions. With the premise that the electrician going to properly select, install, and correctly apply the VFD, this up-front homework is mandatory. A sample form for analyzing a motor and its load is shown in Figure 7–1. This is data that must be gathered to size and select the VFD.

Ambient Conditions

Two of the first questions to ask and address are: Where will the physical VFD be installed? and What are the ambient conditions? As has been shown, location of the VFD is important. It should be as close as possible to the motor it is controlling. Next, the VFD will most likely have to be installed within a protective enclosure. That determination is made based on the conditions of the surroundings. Is the atmosphere dirty or clean? What are the actual temperatures, at all times of the year, around the proposed drive installation? Conditions that are too hot or too cold demand immediate consideration whenever selecting the drive, enclosure, and air-conditioning requirements. If the ambient conditions are hot, then our VFD has to overcome that condition as well because the VFD generates its own heat. Are the surroundings prone to high humidity or moisture? Moisture will condense, and if it condenses on the VFD, then a possible conductive path can be present, especially on a sensitive, low-voltage control board (short circuit). Are any corrosive elements in the air? Is the area hazardous? What is the altitude of the installation? Heat does not dissipate as well at higher altitudes.

Asking these questions first, instead of selecting and installing the VFD *and then finding out there is problem,* makes a lot of sense. All of the aforementioned ambient installation concerns relate to using a little common sense. With these concerns addressed, better selection of an enclosure for the VFD can be made and, if need be, a VFD that may have

Motor Nameplate and Field-Test-Data Form

Employee Name _____

Company _____

Date _____

Facility/Location _____

Department _____

Process _____

General Data

Serving Electric Utility_____

Energy Rate ($/kWh) _____

Monthly Demand Charge ($/kW/mo) _____

Application _____
Type of equipment that motor drives

Coupling Type_____

Motor Type (Design A,B,C,D
AC,DC, etc.) _____

Motor Purchase Date / Age _____

Rewound _____ ❏ Yes ❏ No

Motor Nameplate Data

1. Manufacturer _____

2. Motor ID Number _____

3. Model _____

4. Serial Number _____

5. NEMA Design Type _____

6. Size (HP) _____

7. Enclosure Type _____

8. Synchronous Speed (rpm) _____

9. Full-Load Speed (rpm) _____

10. Voltage Rating _____

11. Frame Designation _____

12. Full-Load Amperage _____

13. Full-Load Power Factor (%) _____

14. Full-Load Efficiency (%) _____

15. Service Factor Rating _____

16. Temperature Rise _____

17. Insulation Class _____

18. kVA Code _____

Motor Operating Profile

	Weekdays Days/Year __	Wknd/Holiday Days/Year __
Hours	1st Shift _____	_____
Per	2d Shift _____	_____
Day	3d Shift _____	_____

Annual Operating Time _____ Hours/Year

Type of Load (Place an "X" by the most appropriate type.)

_____ 1. Load is quite steady, motor "On" during shift

_____ 2. Load starts, stops, but is constant when "On"

_____ 3. Load starts, stops, and fluctuates when "On"

Answer the following only if #2 or #3 above was selected:

% of time load is "On" _____%

Answer the following only if #3 was selected:

Estimate average load as a % of motor size ___ %

Measured Data

Supply Voltage
By Voltmeter

Line $\quad$ V_{ab} _____
to $\quad$ V_{bc} _____ $\quad$ V_{avg} _____
Line $\quad$ V_{ca} _____

Input Amps
By Ampmeter

A_a _____
A_b _____ $\quad$ A_{avg} _____
A_c _____

Power Factor (PF) _____

Input Power (kW) _____
If available. Otherwise equal to:

$$V_{avg} \times A_{avg} \times PF \times \sqrt{3} / 1000$$

Motor Operating Speed _____
By Tachometer

Driven Equipment Operating Speed _____

Figure 7–1 Motor nameplate and field-test-data form. (Reprinted from the U.S. Department of Energy's Office of Industrial Technologies BestPractices reference materials. Call the Information Clearinghouse at 1-800-862-2086 or visit the website, http://www.oit.doe.gov, for additional information.)

to be oversized for the application just to handle the heat load can be chosen. After all, the VFD is going to be installed to deliver power to a motor to drive a particular load. This loading will directly relate to the amount of heat that will be generated by the VFD. Dealing with that heat is critical.

Constant versus Variable Torque

In selecting a VFD, the type of loading to the motor *must* be considered Is the load variable torque or constant torque? Torque, which was discussed in Chapter 2, is what moves the objects, material, and loads in any machine or system. Horsepower does not really do the work. Logically, if there is any work to do whatsoever, then there will be the need for torque. This is true, but what amount of torque at that speed is required? Using the curves in Figure 7–2, the two types of torque can be segregated. As can be seen in the curve, variable-torque applications require very little torque at starting and at low speeds. These are often centrifugal loads, which do not develop the need for high torque until their speed has increased dramatically. These are the "easier" applications for AC drives to operate, and these are the applications where true, exceptional energy savings are justified. While it is true that some torque is required anywhere on the speed curve, the full, high-load torque seen in constant-torque applications is not required. The safest approach to applying the AC drive is to know the load requirements at all possible operating speeds. Select the drive based on the worst case. Some engineers feel that if they cannot turn the motor shaft by hand (indicating a light load to start) then the AC drive may have trouble (this approach is very conservative). Many fans and pumps are variable-torque, centrifugal applications and are good candidates for AC VFDs.

Constant-torque applications on the other hand, have to be analyzed well. These are the ones that can possibly get the VFD into trouble. How much torque is required and when? How low in speed will the drive have to operate with that load? These are questions that must be answered first in constant or continuous high-torque applications. These applications will usually have high-starting-torque requirements as well. The AC drive can be sized to handle these types of loads, but up-front investigation can save embarrassment and hassles later. These applications are not justified for energy-saving reasons but rather for better process control or soft starting. The AC drives today—with higher carrier frequencies and with vector capabilities—can handle most constant-torque applications.

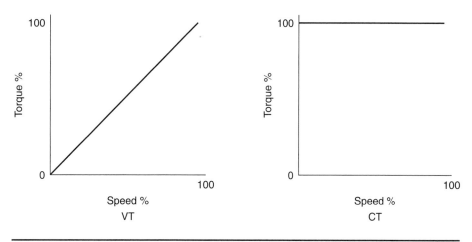

Figure 7–2   Comparing variable-torque (VT) loads with constant-torque (CT) loads.

High-Starting-Torque Applications

Those applications that have that constant-torque look have to be checked over completely before selecting a VFD. The load conditions in these cases demand that the AC motor be given higher currents at start to develop the torque required to start moving. Typically, a NEMA design B motor will provide 200 percent torque whenever it is given 600 percent to 700 percent of its full-load-current rating. However, when a VFD is placed in the circuit between the line and the motor, it is for the purpose of being a current limiter. In normal operation, the VFD is monitoring the current out to the motor so that any instantaneous levels are detected and the VFD shuts down with a fault of overcurrent (its function being to protect the motor). The drive is merely doing its job, but unfortunately the application is not satisfied. When this application was line started, there was enough current allowed to go out to the motor to start the load; now the presence of the VFD is keeping that from happening. This is where it may be necessary to select a VFD of a higher current rating than the motor's rating. This is done to ride through those instantaneous surges to the motor required at start.

Another function of the VFD-related output current is the drive's current-limit setting. This is an adjustable parameter that can help both with the high-starting-torque requirements of a load and also acting as an *electronic shear pin* for sensitive loads. If ride-through capability is needed in starting loads, the current limit can be adjusted up over 100 percent, but most drives do not permit much over 10 percent to 15 percent rated. This still means that the drive may have to be replaced with one of a higher rating to start a high-current-demand motor. However, the current-limit setting can be adjusted very low—virtually to zero—and the drive, upon meeting this level of output current, will trip. In this manner, the load and motor are protected.

Multiple Motors

Two or more motors in a process often will have to run the same speed. In these applications, instead of installing an individual solid-state, variable-speed drive for each motor, a single drive of large enough capacity can be utilized. For example, many supply-and-return fan systems have to run nearly the same speed. Maybe the sizes of the motors match and maybe they do not, but a single drive sized accordingly can handle the total current draw of both motors. If a 20-horsepower, 460-volt, three-phase AC motor with a full-load-current rating of 26 amperes is used on the supply fan and a 15-horsepower, 460-volt, three-phase AC motor with a full-load-current rating of 21 amperes is used on the return fan, then, by adding the total full-load amperes, *26 A + 21 A = 47 A* total is required. Select a variable-speed drive that can handle the full-load total amperes of the motors that it must control.

As can be seen in Figure 7–3, when running two or more motors off of one drive, it is necessary to place individual thermal overloads in the circuit for *each* motor. This is done so that the 47-ampere drive does not inadvertently try to send a full 47 amperes to either of these smaller motors in a case where one is out of service. The drive's electronic-thermal-overload circuit does not distinguish the quantity of motors; it is only monitoring total electric current over a given period of time.

Closed-Loop and High-Performance Drives

Variable-frequency drives today are required in many applications to deliver performances in speed and torque regulation that exceed the standard designs of the drives. This means that the *base* drive will have to be equipped with special software algorithms, regulators, and modules that can accept feedback from the process or system. These are those applications that require very exacting speed and torque control. To give the VFD controller the

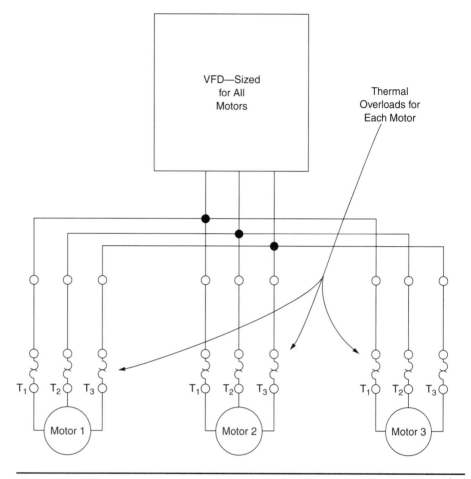

Figure 7–3 Running multiple motors from a single VFD's output (all run at the same speed).

information necessary to meet these requirements, feedback devices have to be implemented into the system. Many AC drives are turned into high-performance, flux-vector drives to deliver the speed and torque needed. Additionally, any positioning drive system must utilize feedback to close the position loop. Ironically, these feedback devices are not going to be the most expensive pieces of the system but may loom as the most important.

To better understand higher performance motor speed and torque control, it is first necessary to review open-loop control versus closed-loop control. Open-loop control, in its most basic form, does not lend itself well to fully automating anything. An input signal or command is given, and a direct resultant output is achieved. The operators do not affect the signals; nor do they know if what they requested actually happened. It might be safe to assume that, if an open-loop control scheme is at work, there are no sensors or feedback devices present in that system.

A good example of feedback is the human operator. The questions often arises, "How is the machine being speed controlled now?" The answer comes back, "Our operator goes over to the machine and turns a speed potentiometer a half turn or so and the machine is running now where we like it!" The operator is the feedback device as he or she is continually correcting the system. If the operator forgets to adjust the potentiometer, then the machine runs as *open loop*. Wherever it is commanded to be, the speed remains, unaware that there may be a speed-related problem. This is a prime candidate for the implementation of some automatic sensor or feedback device to close the loop and automate the speed-regulation loop.

An *open-loop system* is shown in Figure 7–4. It is basically a process that is commanded to do something and the signal going out to the process to accomplish this task is sent out. The process may or may not do as is intended. The *closed-loop system* can be seen in Figure 7–5. Note that there is another component in the diagram, which is providing information about the process. This information returns to a summing or comparison point for analysis, and then new output is provided based on the comparison with the process input signal or the "what should we be doing?" signal. This is a cause-and-effect relationship between devices. The feedback device tells of any actual error in holding to the prescribed condition, and the controller makes the attempt to correct for that error. A closed-loop control scheme is constantly correcting itself. A good example of a closed-loop system is that of a cruise-control function in the automobile. A speed-sensing device located about the car's wheels lets the driver know how fast the car is traveling, and that information is fed back to a control point to compare to the desired miles per hour. The more often the error signal is sampled (sent, received, and corrected to), the better the overall system's accuracy.

With VFDs, there are many types that do require feedback to function at all. Others are given a speed command, and it sends a corresponding frequency and voltage out to the motor. There is a pseudo–feedback loop in that the counter-electromotive force from the motor is being monitored against the speed command and there is some correction for

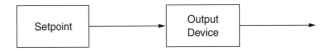

Figure 7–4 Open-loop control scheme.

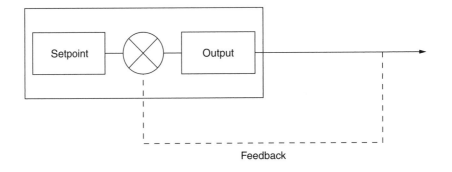

Figure 7–5 Closed-loop control scheme.

speed error. However, this type of drive is not considered a closed-loop system. Once a pulse generator is attached to an AC motor and accommodations are made at the AC drive to accept that feedback, then we have a closed-loop AC drive (some form of AC vector drive). Likewise with the DC motor, there is usually a tachometer to provide armature-speed feedback or even an encoder added to the DC motor to provide more exact feedback to control motor speed better. As for servo systems, feedback is the essence of these applications. The term *servo* implies closing a loop in some way. Either controlling the speed extremely well or positioning a motor shaft, the servo drive—and even stepper drives—depends on feedback.

Most motion usually involves a rotating element or electric motor. To this rotating element are attached the feedback devices, most often directly connected or coupled to a motor shaft. As the shaft spins, the sensing device gathers speed or position data or both. Prior to feedback devices in the electric-motor-control arena, there were first attempts to control the motors with the electric data available. Alternating current motors can provide a form of feedback in the electric circuit called *counter-electromotive force,* or counter-EMF, or cEMF. This is a form of voltage that the VFD monitors. The drive takes this voltage through on-board devices called *potential transformers (PTs),* and it is converted into a usable form for the controller's logic section. Likewise, in current-sourcing drives there are *current transformers (CTs),* which provide a current reading from the wire running through them. This current value is also used in the drive's control-logic section. This electric data can be used by a motor controller to determine how fast a motor might be running—with some inaccuracies—and thus provide correction to the output signal to the motor. Depending on how well the speed regulation has to be for the given motor and application, this means of feedback can be adequate.

Besides current transformers, other methods utilized by VFDs to sense current are Hall-effect sensors, AC shunts, and DC shunts. Direct-current-resistive shunts are very similar to AC-resistive shunts but are mainly used in DC motor and drive systems. They both are low-cost methods of measuring current and are easy to understand, as they merely apply Ohm's law. They require no external power to operate and are very reliable, and there is no zero offset (at zero current there is zero output). They are used from time to time in drive products but are not electrically isolated, which makes them susceptible to electric noise. The output commonly has to be amplified to be used in the control circuit, and, for this reason, it is not widely used. The Hall-effect sensor is a magnetic core with a coil that surrounds the wire. The Hall generator measures the value of current, and then this signal is amplified. A separate power supply is required, unlike the current transformer, which does not need one. Hall-effect devices and current transformers (CTs) often are confused with one another in appearance. They are different in that the CT can only measure AC current, whereas the Hall-effect device can measure both AC and DC current. The Hall effect is similar to the current transformer (CT) in that it is reliable and electrically isolated. The Hall-effect device is used to measure current within the VFD.

Braking Methods

One often-confused subject in motor control and in solid-state VFD applications is that of braking and regeneration. How does it work when a VFD is used? What type of VFD should be used? Does the drive have regenerative (regen) capabilities, and is dynamic braking also required? Which method is the fastest at stopping the motor? Which is the safest? Does the application even require braking? Should a mechanical brake be used? These are all good questions—some with more than one answer.

An electric motor, whether DC or AC powered, moves its desired load and demands a certain amount of power to do its job. However, when the load decides to drive the motor then the motor is a generator—a generator of power. Where does this energy go? The energy coming from the motor is counter-EMF—also known as back-EMF. The magnitude of this EMF is the main concern in braking or regenerative situations. If this cEMF has a channel to get fully back to the source of supply power, then it is called *regenerative.* If this channel is used to slow or stop a motor and its associated load, then it is called *regenerative braking.* A high-inertia, low-friction load; an overhauling load; or a crane or hoist, are all applications in need of stopping. Thus, the choices are controlled stop by ramping, dynamic braking of the motor, DC injection braking, or regenerative braking of the motor. A mechanical brake should only be considered for E-stop (fail-safe) and hold situations.

Regenerative braking assumes that there is a means of getting the load-generated energy back to the supply mains. It also assumes that this means or channel is operating. If a current-source AC drive or a regenerative DC drive has faulted and is no longer in control of the motor because it has lost its control power, then regeneration through its power bridges via the firing of SCRs is not going to happen. Thus, it is usually customary to include dynamic braking as a safeguard in those applications needing ensured stops. This involves adding a circuit (contact activated), which, upon power loss, will dissipate the motor-generated energy to a resistor bank. This will stop the motor fast.

Dynamic braking takes mechanical energy that is being backfed through the system as electric energy and dissipates it as heat at a resistor bank. Voltage-source drives, diode-rectified drives, and PWM inverters utilize this approach to braking an electric motor. This method of dissipating regenerative energy is also used to slow a motor so as to provide back tension or holdback torque as in winding and unwinding applications. This is, of course, used when regeneration back to the mains is not possible. It should be noted that when a drive that is not fully regenerative tries to control a motor in a generating state, the DC bus voltage rises quickly and an overvoltage fault will occur; if the regeneration rate is too dramatic, then power devices can actually explode! Obviously, dynamic braking wastes energy as heat. Therefore, if the application brakes often and quickly and the loading is heavy, then alternate methods should be explored. The *regenerative drive* is one. It allows this electric energy to go back to the mains. This type of drive must have the appropriate number of power semiconductors and the right type to accomplish this. There is a premium for this extra hardware, and it must be decided if the energy losses in heat outweigh the up-front costs of the hardware in a regenerative system.

Another method used as an alternative to regenerative baking is that of *common bussing.* In this scheme, regenerative power can be used to power another motor that is in a motoring state. This power is resident on the bus for the appropriate use, and this scheme is sometimes called *common bus.* It can only be used in those instances where the machine has a motoring component and a generating component. This can typically be found on a line that has an unwind motor and a rewind motor. Many DC systems utilize plugging to accomplish braking. This is hard on the DC motor armatures (full reversal of the polarity), the AC/DC reversing contacts necessary to accomplish this, and all the associated mechanical parts in the drivetrain.

As far as the electric motor is concerned, it does not matter if braking is accomplished via resistors or through regeneration. The motor has become a generator, and that energy is going somewhere. The VFD must have the necessary logic and chopper circuit to divert this energy. Today's dynamic braking utilizes solid-state components and drive-integrated logic to ensure proper sequencing. The conventional contact to remove the armature power supply in a DC system can now be replaced with an SCR. The goal is to take this motor-generated current and reuse it as braking torque. *Deceleration* is a form of regenerating motor energy to achieve slowing and stopping. Deceleration can be described as controlled stopping. It is common to see a holding brake electrically actuated when there is no motion at the motor. Once in normal running mode, these types of brakes should not be used, so the cEMF dynamic braking through resistors or back to the mains can be optimized.

Sizing of dynamic-braking resistors usually is left to the drive manufacturer. However, it is easily calculated if one knows the following:

> Duty cycle—braking times and how often
> Maximum speeds
> Power rating of motor and efficiency
> AC drive DC bus voltage

Another form of braking is *DC injection braking*, sometimes called simply *DC braking*. A DC voltage is forced between two of the phases in an AC motor for a given period of time causing a magnetic braking effect in the stator. This type of braking is commonly found in systems that can tolerate braking at lower speeds. The reason for not applying this type of braking at higher frequencies is that the brake energy can remain in the motor and could quickly cause overheating.

Other Considerations

Other considerations to factor in whenever applying VFDs on AC motors include:

- Soft-start, ramping-to-speed, or S-curve capability. Does the load need this?

- Power supply, common input voltages.

- Speed range. How slow is *slow* in motor rpms? Check the motor.

- Existing AC motor retrofits. Check the motor's nameplate and get a drive to match its characteristics.

- Single-phase and three-phase power. Three-phase drives can run with single-phase input power but must be derated. Consult VFD manufacturer.

- Does application need variable-speed capability? If not, then why apply a VFD?

- Is application outdoors? If so, this environment is going to be difficult for a VFD to operate in as it needs cooling.

- What are the horsepower and current requirements? There are many different types of drives on the market. Some have more complexity than others. Pay for only the capability that is needed. The current rating and voltage rating of the drive often dictate the type and physical size.

- Make sure a drive physically will fit in the desired enclosure, room, or space planned. Do not be surprised after it arrives and will not fit through a door to get into a room.

- Power factor. Today's VFDs have unity power factor, which is typically a high rating. This will help a facility's overall power-factor concerns.

CHAPTER

8

Industrial and Commercial Variable-Frequency-Drive Applications

An AC VFD can be made to control almost any polyphase AC induction motor out there. That makes up millions of motors, and many of these motors are running such that a good deal of energy is being wasted. In other instances, machines and processes are limited to one constant speed and this is not acceptable. These applications are prime targets for the implementation of the VFD. Using much of the base knowledge presented in this book, the VFD installation should result in success. It should be noted, however, that a VFD does not make up for application and system inadequacies. For instance, it cannot correct for a motor that is undersized for its load or show tremendous energy savings in a heating, ventilating, and air-conditioning (HVAC) system that is not properly calibrated. A VFD does what it does very well—changes the speed of an AC motor as commanded. It does not fix external problems.

Heating, Ventilation, and Air-Conditioning Systems

Variable-frequency drives are used to maximize the performance of many fan and pump systems in facilities all over the world. By adhering to the fan and pump laws, the VFD can provide energy savings, especially where energy costs are high. A fan or pump's horsepower requirement is reduced by a cubic function of the flow rate. Thus, a minor reduction in air or water flow will yield a dramatic reduction in power. A major problem with HVAC traditional motor-driven systems for air or water-flow control is oversize. The motor horsepowers are greater than they need to be, and, thus, dampers and inlet guide vanes (fans) or discharge and flow-control valves (pumps) are employed to get the desired flows. This is a terrible waste of energy. Just incorporating a VFD into the application does not guarantee savings. Actual sizing of motors for their given loads should be considered along with an analysis of control—tuning and calibration of the overall system. An untuned HVAC system with VFD equipment can actually use more energy, especially if motors are still oversized! One initial problem with an oversized HVAC system is that any associated new equipment must match the oversized equipment. A VFD placed on an existing fan motor must match the motor ampere rating. If the motor is too big for the fan and duct system, then there is an inefficient use of energy.

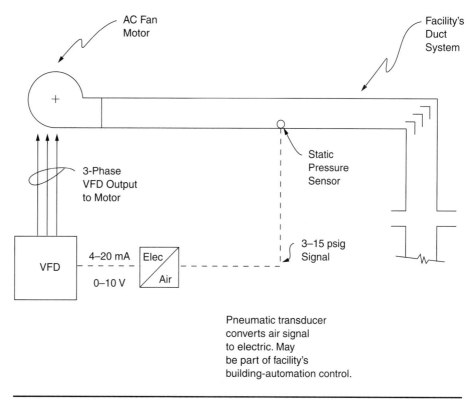

Figure 8–1 A simplified VAV system utilizing a VFD to maintain a level of static pressure.

One common application of VFDs in a facility's HVAC environment is in a variable-air-volume system, commonly referred to as a VAV system or static-pressure-reset system (Figure 8–1). These systems typically work on maintaining a static pressure level in the ductwork. If the static pressure in the duct rises above the setpoint, then the VFD should lower the fan speed; if the static pressure goes below the setpoint, then the fan motor speed must be increased. Most VAV systems try to maintain static pressure levels of 1 to 3 inches of water pressure and are designed around a full-load condition. During periods of lighter load on the air system, there may be different static-pressure setpoints to regulate to. Again, the VFD is able to continually adjust the speed of the fan motor rather than using an air-actuated or electrically actuated damper system while a constant-speed motor is consuming full energy. Typical fan curves are shown in the next chapter.

Variable-frequency drives placed into VAV systems for energy systems can fail to show savings for other reasons than just oversized and poorly tuned applications. If a designer believes that the setpoint should not change, then the system will not perform under *all* conditions, that is, other than full-load design. There are many different temperature fluctuations with the weather, occupancy considerations, holiday and evening settings, and light-load-design considerations. There must be multiple setpoints built into the controls. Another factor is that of pressure-dependent and pressure-independent terminal boxes.

These boxes work with static pressure reset, and a static pressure point must be found to maintain comfort in any particular zone. Find these comfort zones, and design the VFD/VAV system around them.

Another benefit provided by the VFD installation is the newfound ability to track energy and power usage. Many of today's VFDs have built-in kilowatt meters that tally the energy usage and can even send this data out to a remote data-collection system. In this way, adverse conditions in the HVAC system can be tracked: dirty or clogged filters, dirty coils, and so on. Preventive maintenance can be scheduled based on this data. Lastly, the VFD allows for flexibility. Any changes to the system can be experimented with first by trial-and-error activities at the VFD and motor.

Many of these VAV systems are incorporated into existing supply and return fans. Since the volume of air being supplied and that being returned is close, the size of the motors can be just as close (if not exact) and the fan speeds can be matched. This sometimes allows for one larger VFD to be used to control two smaller fan motors. This can help keep the initial cost down. Also, the facility managers must determine whether or not a need for running at full speed will be required in the event that the VFD is faulted or in need of repair. These are constant-speed bypass systems, and care must be taken whenever the bypass system is employed because ductwork can be damaged if the dampers are not open to the point that full 60-hertz motor operation is expected.

Another application of VFDs in facilities is for cooling-tower fan control based on temperature. As is seen in Figure 8–2, a cooling-tower fan motor is controlled by a VFD to

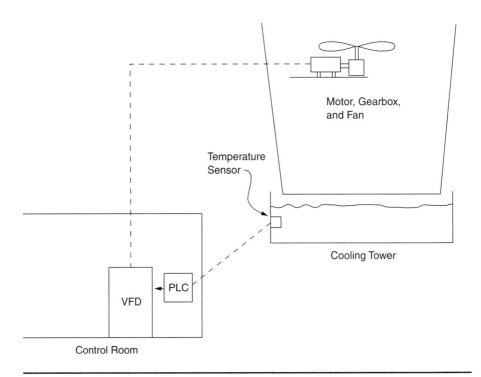

Figure 8–2 Cooling-tower fan motor controlled by VFD.

maintain a setpoint of temperature. Outdoor temperature conditions change daily, and, thus, a motor will have to run faster or slower to keep the water at a desired temperature. One issue with VFD installations on cooling towers is the cable distance from VFD to motor. Most often, the motor is several feet above ground level at the top of the cooling tower. This means that the VFD, which cannot be located near the motor, is potentially well over 100 feet away. Attention should be given to the possible voltage spikes that can be present in these instances.

Other applications (not necessarily for heating or cooling) can include the use of VFDs on the following: chilled-water pumps, condensate pumps, freshwater pumps, effluent and wastewater pumps, some compressors, and boiler fans. In many of these cases, the VFDs are not employed for substantial energy savings but rather for performance issues and not stressing the motor and mechanics of the system by utilizing the VFD's soft-start characteristics.

Variable-frequency drives have made strides into many applications that traditionally were handled by DC equipment. Many processes and machines can be VFD controlled with either the volts-per-hertz drive operation or going to the higher-performance control scheme of a VFD's flux-vector operation. In doing so, applications such as printing presses, extruders, paper machines, elevators, cranes and hoists, and web-handling processes can sufficiently run utilizing AC drives. A typical web-handling system depicts a process whose VFD use is for tension control of the web, the material being pulled through a process by the motors. Some systems run in a *draw mode,* which runs a particular motor just a little faster than another. In this way, some tension is controlled, but utilizing a drive's current regulator and controlling the torque is the best method for these applications.

These tension- or torque-control applications call for regenerative drives capable of holdback torque. Industries such as paper, film, foil, packaging, winding/rewinding, and web converting all demand this requirement from the drives. This makes these applications predominantly DC, although AC VFDs are used, especially the flux-vector types. The web-converting line has to coordinate the activities of many motors and other machine functions. The line or machine is converting a web of material as it unwinds a roll of raw material. The web can be slit into narrower strips, coated and dried, laminated to another web, painted or printed on, and then wound into a final roll. There are zones between certain sections where tension must be controlled. These are called *tension zones.* The main control method is tension control. The tension between sections of the machine has to be not too tight (material can break apart) and not too loose (material could droop, sag, or drag). This can be accomplished by precisely speeding up a downstream motor or by maintaining better control of torque at each motor. In these systems, torque equates to tension. A line such as this needs to have high-speed, high-response control equipment running the motors, electronic drives, and other machine controls. If not, material or webs break or they unroll all over the place, making a mess. The goal is to run the process at the fastest speeds possible. Thus, controllers have to coordinate between devices, sometimes down to the millisecond range for electronic and processing response. Many sections in the web-converting line will derive transducer feedback from the line to use in controlling tension. This can be in the form of voltage from a dancer, load cell, strain gauge, linear-variable differential transformer (LVDT), or ultrasonic device. This signal is fed into an electronic drive or other master controller for use in high-speed processing. Once processed, the tension feedback is now in the form of a new torque output to the motor. Additionally, line speed and overall line tension are set by a local operator, via a remote operator station, or by a master controller.

Proportional, Integral, Derivative Loop Control

One of the most predominant control schemes used throughout industrial and commercial facilities is the proportional, integral, derivative (PID) loop. Also very common is PID motor control in industry utilizing VFDs. The programmable-logic controller (PLD) has often been called upon to perform more extensive calculations to control the process. In many chemical plants and in water and wastewater facilities, a distributed-control-system (DCS) scheme is often utilized. In many office and school buildings, yet another system—the building-automation controller (BAC)—is employed. All function in an overall similar manner: monitoring, manipulating, and turning input/output on and off to control a process or system. However, VFDs have gotten more sophisticated and have been asked to deliver this performance. The same functionality is asked of the VFD system. If the VFD controller can handle the extra requirement within, maybe by adding a module or two, then other devices (another motor, bigger tank, longer dryer, more sensors, etc.) will not have to be added. If a sensor can be added and the feedback directed into the VFD for evaluation, then the process might be more accurately controlled and, thus, more elaborate and expensive modifications to the process can be averted. It is still true that the fewer the components in a system, the more reliable that system is and the more interconnecting cable and wire between multiple devices, the more likely that data transmission and signal errors can occur.

The VFD can be used as a setpoint or PID-loop controller. As a matter of fact, today's VFD-based systems are expected to have PID capability and the software and predefined function blocks to perform it and must be able to have the flexibility to perform variations of the algorithm. There are simple P loops (full proportional)—flow needs changed, therefore change the speed. There are PI control loops. There are interactive and noninteractive PID loops. Some are adaptive, which means the controller can make additional corrections—above and beyond the set PID-loop changes—as the process warrants. These can involve gain changes to the overall system. The *gain* of any system is defined as the ratio of the system's output signal to its input signal and can be affected by tuning a system to be more or less responsive. Figure 8–3 shows practical PID loops graphically. The PLC is acting as the "traffic cop" in the system. It is monitoring temperature and turns the heater on and off. It also is watching the level of the tank and controls the flow of the liquid into the tank. These are full, closed-loop, process-control systems utilizing the PID scheme. While full discussion on PID loop control is beyond the scope of this book, its use with VFDs is very common.

A PID loop—being a closed-loop system that receives feedback from some device and compares that feedback with a reference or setpoint for the process that it is controlling—is made up of three distinct components. Often times, all three components of the PID loop are not incorporated in the actual control scheme. For example, a control-signal output to a device that is proportional to the sum of the error signal and the setpoint is the *P* portion of PID. The control signal is directly correcting for any deviation seen in the process. This is often called *PID control,* but, in actuality, it is simply *P* or *proportional* control. For any given action, there is a proportional reaction by the system's controller.

Likewise, the *I* portion of the loop control is the *integral* of the error signal with respect to the setpoint. This is where a good understanding of calculus comes into play. Integrals take into account time. Resetting values through a corrective control loop, during a period of time, takes place in this type of control. This looping is dependent on the severity of the deviation to the setpoint and the processor's capability. Some systems can employ both the P and the I portions of the PID control scheme.

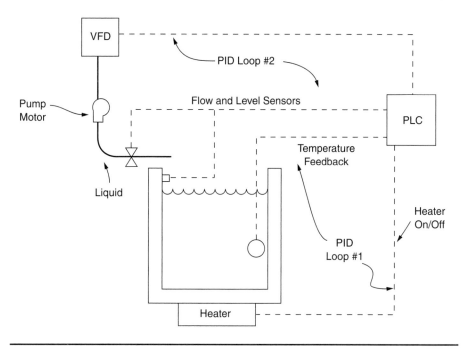

Figure 8–3 Setpoint or PID control of a process.

Lastly, the *D* portion of the PID loop is the *derivative*. A derivative is a change to, or derivation, of the resultant. This loop is more concerned with rate of change in the deviation. Similar in some ways to the integral loop, controller scanning and how often the deviation is seen become major components of the equation. This method of the PID loop can be coupled with only the P portion when the application dictates and is referred to as a PD control loop.

Proportional, integral, derivative control can be as simple as shutting a valve off when a desired temperature is met or as complicated as instantaneously correcting a machine's motor speed, inertia, and torque as a material builds on a roll at high speeds. An important facet of PID control is how fast in milliseconds, or even microseconds, the entire loop takes. This is a function of the microprocessor speed, where data is coming from (and how fast the transmission is), and how well the program and calculation methodology is laid out. Sometimes high-frequency filters are needed to keep PID signals true (integrity sound, low electrical noise), because any deviations to these signals can be perceived as major fluctuations in a sensitive, high-response, and high-resolution control scheme. By doing PID functions right at the VFD, valuable processing and transmission time can be saved.

The *transducer* is a common feedback device used in gauging, measuring, and detecting displacement and is used in many PID/setpoint controlled systems. A transducer is a device that converts mechanical energy into electric output. A better, more appropriate definition is that variations of this device can convert any input energy, including electric, into output energy. The output energy will, in turn, be different from the known input energy, thus supplying useful feedback information from which closed-loop control and correction can be achieved. There are many different kinds of transducers, and there are just as many ways to provide the input and output to the transducers. The most common include pneumatic, hydraulic, electric, LVDT, pulsing, and ultrasonic types.

Transducers sending an electric output signal sometimes send that signal as a voltage, 0 to 10 volts DC, and sometimes as current, commonly 4 to 20 milliamperes, with 4 milliamperes being an *off* condition, 20 milliamperes being full *on,* and any value in between being the range of the transducer. This range is generated from a stimulus to the transducer and corresponds to some direct action from which the tranducer's output signal is going. For instance, a process involving the unwinding and rewinding of a material utilizes a *linear-displacement transducer,* sometimes called a *dancer.* Changes in the dancer position correspond to a voltage drop across a resistor. This voltage-output value from the transducer indicates how loose or tight the web material is and thus sets the span. From here, a correction can be made by the motor drive controllers to increase or decrease speeds and/or torques.

The *piezoelectric transducer* is another type of feedback device that produces motion from an electric stimulus or produces an electric signal from the motion. These types of devices can be called *strain gauges* and *accelerometers.* Another device is the rotary-variable differential transformer (RVDT). This device is a unit employing a rotor-and-stator relationship to produce voltage. No slip rings are used as the electric output is via the electromagnetic relationship of the stator windings and the rotor. It produces a voltage, usually 0 to 10 volts DC, whose range varies linearly with the angular position of the shaft.

Process Control with Sensors

In any process-control situation, there is the need to control one or many conditions: speed, flow, temperature, pressure, tension, and more. There are many different types of sensors and transducers out there, and their proper operation and feedback into the VFD dictates how well the closed-loop control scheme will work. Sometimes the temperature can attain extremely high levels; thus, an industrial grade of temperature sensors has emerged. The class of temperature-sensing devices used that generate electric output are called thermoelectric devices. One common device is the *thermocouple,* which utilizes two metallic components and will conduct an electric signal when their junctions are at different temperatures. It is sometimes referred to as a *thermal-junction device.* Other temperature-sensing devices use a direct output linear to the measured temperature whenever feasible. A *thermistor* is an electric-resistive device that relies on its resistance to vary with a change in temperature. A *thermostat* is yet another temperature-sensing device that often utilizes a bimetallic scheme to sense when a predetermined temperature has been met. At this point, a single output signal is provided, thus starting or stopping another piece of equipment; such is the case in a furnace's operation. Temperature-sensing equipment in the factory of today has to be able to interface with many higher-level systems. This means that the simple thermometer does not suffice, and, thus, digital displays, along with microprocessor-based units, have emerged for communicating with other devices and for charting and recording of temperature data in a process.

Additionally, present-day pressure-sensing devices must be industrially hardened and smart—intelligent in the microprocessing sense. They must be rugged for industrial use and have fast action and high precision to be useful. These kinds of devices, much like their temperature-sensing counterparts, get installed virtually in line with the fluids that they are measuring or sensing. Thus, the integrity of the unit is challenged as it physically sits in the wet and hostile environment. Common pressure ranges go up to 500 millibars (mb, the unit for measuring pressure) or 100 pounds per square inch gauge (psig). The accuracy of the pressure sensor is rated by its *linearity*—error from steady-state input/output; its *hysteresis*—the difference in response to an increase or decrease in the input signal; and its *repeatability*—the deviation over many readings, which should be the same. Many times, the pressure sensor converts the pressure signal into an electric signal, and other times this conversion is done in an external controller. This is sometimes an overlooked facet of the pressure-control loop. This conversion module has to physically reside somewhere, and the application and environment will typically decide where that location is. Once located, the conversion is simple, usually taking a 3 to 15 psig or 0 to 100 psig pressure signal (scaling is important) and changing it to a 0 to 10 volts DC or 4 to 20 milliampere electric signal. From here, the controller scales the electric signal further for its use and a resulting output signal is generated.

Automated Systems and Machine Control

The VFD is used to do more than vary the speed of an electric motor. First, in order to run the motor, the VFD must be programmed for the control scheme that it is expected to utilize. Next, the VFD is expected in many automated systems to perform some minor logic operations, determining when and how fast to run the motor it is controlling (preset speeds, speed-coincidence conditions, etc.). Basic VFD control schemes typically are called two-wire and three-wire control schemes. Two-wire control schemes utilize just two wires for the sequencing (run/stop) phase of operation, whereas three-wire schemes need a third wire. These are shown in Figure 8–4. Thus, the starting, stopping, and directional (motor rotation) commands necessary to motor operation are picked up here. All machines with AC motors and VFD control have to have a determination made as to *how these signals are to be sequenced and from where they shall come.* The drive is not all-knowing; the user must tell the drive this basic control scheme.

The *reference* is the term given to that signal that the drive receives, typically from an external potentiometer or PLC controller, that directly relates to how fast it is to make the AC motor run. Reference signals can be in a 0 to 10 volts DC, 0 to 5 volts DC mode or can be 4 to 20 milliamperes. This signal can be scaled inside the drive to correspond to low and high speeds at the motor. There are usually bias adjustments for these signals, as many

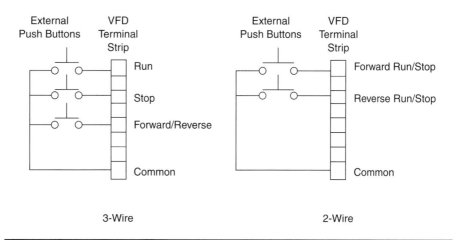

Figure 8–4 Basic two-wire and three-wire VFD control schemes.

times they are not true and exact voltages. This flexibility by the VFD allows for deviations to be corrected. Additional control terminals at the drive are low-voltage inputs and outputs, with many being configurable by the user for the machine or process. There are contacts to indicate that the VFD is running OK and that there are no faults. Additionally, many drives come equipped to send analog signals out to drive-speed meters. Thus, it is possible to utilize many control options whenever setting up a VFD on any one machine. The VFDs, equipped with extensive and programmable capabilities, can be made to be very different from one another when the machine or automated-process needs have to be met.

CHAPTER

9

Energy Savings with Variable-Frequency Drives

Determining Energy-Saving Opportunities

It is widely agreed that VFDs are a tremendous source of energy savings when applied to the proper loads, and much of this chapter will help to confirm that. However, to aggressively reduce energy costs within any facility, it is necessary to understand where the energy dollars are going. This means that instead of complaining about the monthly electric bill and then paying it, why not analyze it, understand it, and then do something about the "problem areas"? There are several areas where electric costs are not just limited to consumption. There are demand charges, service charges, power-factor penalties, load and rate factors, taxes, and so on. Figure 9–1 shows an example electric bill, and Figure 9–2 gives the descriptions of that sample electric bill's components. Interpreting the utility charges is the first step toward applying VFDs for energy savings.

Next, a breakdown of the areas in the facility that use that electricity must be developed. In Figure 9–3 is shown a sample table of a sawmill's energy-end-use summary. Note the motors that contribute the greatest energy needs from the table. Now we have a handle on how much we are consuming and when (from the bill), and we know where the usage is within the facility. Comparing this information with a past year's history may start to yield clues as to where those problem areas are. Once identified, an in-depth study for those particular applications can be made to determine if the application of a VFD will save energy.

With this energy data gathered, it can be determined if demand-management measures can be taken. This is done by calculating the load factor. *Load factor* is the ratio of a facility's average to peak demand and gives an indication of how effectively demand is allocated. Figure 9–4 gives an example for determining a facility's load factor. If the load factor varies significantly from billing period to billing period, then there may be some definite energy-saving opportunities. If the annual load factor is 80 percent or less, then somewhere in the plant demand-reduction measures can be taken. Typically, if the load factor is over 80 percent, then there are probably very few demand-limiting measures to take.

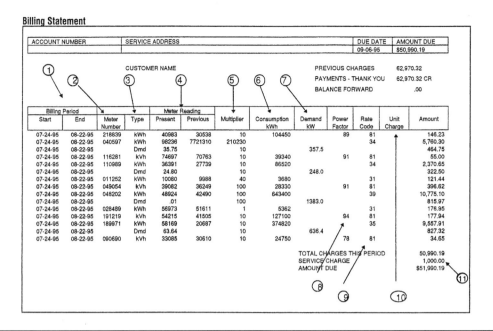

Figure 9–1 Sample electric bill. (Reprinted from the U.S. Department of Energy's Office of Industrial Technologies BestPractices reference materials. Call the Information Clearinghouse at 1-800-862-2086 or visit the website, http://www.oit.doe.gov, for additional information.)

1. Service days—The number of days in the billing cycle.
2. Meter number—The number shown on the face of the meter.
3. Meter type—You could have one or more of the following types of meters:
   A. Energy and demand—This measures kWh (kilowatt-hours) and kW (kilowatt).
   B. Reactive energy only—This measures kVARh (kilovolt-ampere-hour reactance), which is used to bill for power factor less than 95%.
   C. Energy, demand, and power factor—This meter has the capability to measure all three.
4. Meter reading—The actual reading taken from the meter.
5. Multiplier—This meter multiplier, used in calculating the kW demand, total kWh consumption or total kVARh consumption, is on the front of the meter.
6. Consumption—The actual meter reading multiplied by the meter multiplier in units of kWh or kVARh. Reactive power is the nonworking power caused by magnetizing currents required to operate inductive devices such as transformers, motors, and lighting ballasts and is used as the basis for power-factor charges.
7. Demand—This is the actual kW demand and is calculated by multiplying the kW meter reading by the meter multiplier. The demand kW shown is the highest kW recorded by the meter in any one 15-minute period for the billing period.
8. Power factor—The power-factor % shown on the bill is determined from the kVARh consumption described in #6, along with the *real* (working) power and *apparent* (total) power. The actual charge for power factor below 95% is calculated by multiplying the kVARh consumption by the kVARh rate.
9. Rate code—This is the rate that applies to the meter number shown. For customers with multiple meters, more than one rate schedule may apply.
10. Unit charge—The rate being charged for the rate code shown. If the rate is not shown, you have a time-of-use meter. In this circumstance, a second statement is included with your bill that shows the off-peak and on-peak schedule charges.
11. Service charge—A monthly charge often referred to as the basic, facilities, or customer charge. It is generally stated as a fixed cost based on transformer size.

Figure 9–2 Description of items 1 to 11 on electric bill in Figure 9–1. (Reprinted from the U.S. Department of Energy's Office of Industrial Technologies BestPractices reference materials. Call the Information Clearinghouse at 1-800-862-2086 or visit the website, http://www.oit.doe.gov, for additional information.)

Sawmill Energy-End-Use Summary

Process	Electricity Use, kWh	Percentage of Total Use	Cost
Blowers	484,600	4.8%	$12,115
Chippers	101,600	1.0%	2,540
Air Compressors	1,911,200	19.0%	47,480
Hog	76,700	0.8%	1,917
Hydraulic Motors	857,800	8.5%	21,445
Saw Motors	2,092,000	20.8%	52,300
Planer Motors	132,700	1.3%	3,317
Kiln Fans	2,033,800	20.2%	50,845
Boiler Fans	268,900	2.7%	6,722
Lights	376,400	3.7%	9,410
Misc	1,741,390	17.3%	43,534
Totals	10,077,090	100.0%	$251,925

Sawmill Energy-Consumption Disaggregation

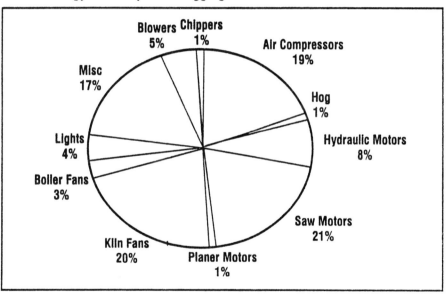

Figure 9–3 Energy usage of a sawmill. (Reprinted from the U.S. Department of Energy's Office of Industrial Technologies BestPractices reference materials. Call the Information Clearinghouse at 1-800-862-2086 or visit the website, http://www.oit.doe.gov, for additional information.)

$$LF = \frac{kWh}{kW_{demand} \times 24 \times N} \times 100\%$$

Where

LF	= Load Factor in %
kWh	= Electric Energy in kWh
kW$_{demand}$	= Electric Demand in kW
N	= Number of Days in Billing Period

Sample Billing Information

Energy Use	Demand	Period
1,132,000 kWh	2880 kW	30 Days

Sample Load-Factor Calculation

$$LF = \frac{1,132,000}{2880 \times 24 \times 30} \times 100\% = 54.6\%$$

Figure 9–4 Determining your load factor. (Reprinted from the U.S. Department of Energy's Office of Industrial Technologies BestPractices reference materials. Call the Information Clearinghouse at 1-800-862-2086 or visit the website, http://www.oit.doe.gov, for additional information.)

Affinity Laws

Alternating current variable-speed drives came into their own in the early 1980s. This was due mainly to the need for saving electric power in the industrial and commercial facilities of the country. By being able to run an electric motor less than its full speed whenever the application did not need full speed, there is the potential for energy savings. In many instances, the energy savings can be substantial. Thus, it did not take too long for owners, property managers, accountants, and practically anybody else interested in a means of saving energy to take notice. Not only could VFDs provide exceptional cost savings while operating, but they could also offer a power-factor improvement, which further makes the utility happy. Alternating current drives limit inrush current to motors. They also provide a buffer (filter) to the utility's supply such that line fluctuations caused by leading and lagging elements on the supply line do not appear as often and are not as severe. Reduced penalties from the utility to the user of the AC drive alone would cause a company to justify the cost of installing an AC drive. However, the real savings is in actual operating costs saved over time. These costs are more dramatic in areas where electric costs are very high.

It makes logical sense: a VFD can be used in place of mechanical restrictors, such as dampers and valves in most centrifugal fan or pump systems. With the fan system, an AC motor traditionally will run full speed all the time and the only way to slow the airflow is to open and close dampers or use inlet guide vanes. This amazingly is still a common method in many commercial buildings even today (and a candidate for applying a VFD). In pump systems, the same reduction in flow was achieved by mechanical means. A pump motor will run at its full-load speed, while discharge or bypass valves are in place in the piping system to be opened and closed—either manually most of the time or automatically—

to reduce the flow of the liquid. These dampers and valves can get stuck, wear out, or corrode. Besides being maintenance items, they help in no way to save energy. The VFD is installed ahead of the AC motor in the electric system. It now reduces the flow electronically by doing the same work but at reduced speeds. It slows the AC motor down and puts out only the voltage and current required to drive the load. With centrifugal applications that follow the same variable-torque load curves, the energy use is dramatically lowered as speed is lowered. These fans and pumps follow certain principles of physics and relationships called the *affinity laws,* which describe the relationships between speed and flow, speed and pressure, and speed and power:

1. Flow is proportional to speed

$$\left(\frac{Q_2}{Q_1}\right) = \left(\frac{N_2}{N_1}\right)$$

2. Pressure^2 is proportional to speed$_1$

$$\left(\frac{P_2}{P_1}\right)^2 = \left(\frac{N_2}{N_1}\right)$$

3. Power^3 is proportional to speed$_1$

$$\left(\frac{HP_2}{HP_1}\right)^3 = \left(\frac{N_2}{N_1}\right)$$

For example, reducing speed to a fan motor to one-half will reduce the power requirements to one-eighth!

These relationships show that the AC drive, when applied to a centrifugal fan or pump, can be a major factor in saving energy over mechanical means, even when running close to full speed (55 to 59 hertz). It will supply full voltage and current if required at full speed, but a centrifugal load does not usually require 100 percent + torque at full speed. These affinity laws, along with other basic formulas for fans and pumps, should be understood by the user, the engineer, or the installer of the drive. This knowledge is useful in getting the full capability out of the VFD. Even before applying the drive, it is sometimes necessary to completely justify the drive from an energy-saving viewpoint. The minimal data required is the cost per kilowatt-hour, the motor size, the hours of operation, and the present means of flow restriction. Once this data is entered into the formula, the calculation is easy. Sometimes the calculation is not even necessary if the facility where the drive is being considered has a very high cost per kilowatt-hour. These are regions where there is high power consumption and the actual production capacity of that power is limited (no new power plants being planned).

As stated earlier, an oversized motor for a centrifugal load is also an energy waster. A VFD can provide remarkable savings, but, if not implemented properly into an oversized application, the energy savings will be nonexistent. Therefore, sizing of these fan and pump systems is critical. Formulas for calculating the brake horsepower (BHP) of a motor follow, and these formulas are necessary for calculating the horsepower in a new installation for fans or pumps (and checking an old installation where existing motors are going to be kept). In existing installations, motor nameplate data is needed along with checking the existing BHP to a recalculated BHP to verify that the motor is not oversized.

Fan Application Formulas

$$Brake\ horsepower\ (BHP) = \frac{cubic\ feet\ per\ minute\ (cfm) \times pounds\ per\ square\ foot\ (psf)}{33,000 \times fan\ efficiency}$$

or, when pounds per square inch (psi) is known:

$$Brake\ horsepower\ (BHP) = \frac{cubic\ feet\ per\ minute\ (cfm) \times pounds\ per\ square\ inch\ (psi)}{229 \times fan\ efficiency}$$

Pump Application Formulas

$$Brake\ horsepower\ (BHP) = \frac{gallons\ per\ minute\ (gpm) \times feet \times specific\ gravity}{3967 \times pump\ efficiency}$$

or, when pounds per square inch (psi) is known:

$$Brake\ horsepower\ (BHP) = \frac{gallons\ per\ minute\ (gpm) \times pounds\ per\ square\ inch\ (psi) \times specific\ gravity}{1713 \times pump\ efficiency}$$

where head in feet is equal to 2.31 pounds per square inch gauge (psig) and the specific gravity of water is equal to 1.0.

Thus, if a prediction were made of what the energy savings would be for a motor whose brake horsepower was determined to be 60 BHP, the first step would be to convert that horsepower value into real power, or watts. Thus 60 times 0.746 (746 watts equals one horsepower) equals 44.76 kilowatts. A VFD is assumed to be at least twice as efficient as a pump with a discharge valve, and, therefore, a ratio is applied to each. Pump systems notoriously operate at 66 percent to 76 percent of their maximum flow rates, and, by using pump curves, which can be supplied by any pump manufacturer, typical efficiencies and other data useful to confirm the ratios can be established. Next, a calculation is made of the power consumed by the pump using the discharge valve; at a ratio of 0.98 to the drive's ratio at 0.5, the calculations are:

$$44.76\ kW \times 0.98 = 43.86\ kW$$

and

$$44.76\ kW \times 0.5 = 22.38\ kW$$

respectively. Subtracting the smaller value from the larger, the difference in energy between the two types of pump motor control is:

$$43.86 - 22.38 = 21.48\ kW$$

This 21.48 kilowatts is the actual power that will be saved every hour. By applying two more factors, time and money (actual hours of operation and cost per kilowatt-hour) a dollar value can be placed on the savings. If the pump runs 8 hours/day, 6 days/week, then in one year it runs 2080 hours per year. If the utility charges 9 cents per kilowatt-hour, then the savings is 0.09 times 2080 times 21.48 kW or $4021 per year. Considering that a 60-horsepower drive may cost $6000 (using the rule of thumb of $100-per-horsepower cost estimate), then saving $4021 each year means that the drive will pay for itself in 18 months.

Owners and managers are looking for payback periods of 18 months or, preferably, quicker. The actual cost of electricity, if high, along with power-factor and demand charges imposed by a utility, virtually ensure a good payback period, especially if the fan or pump runs 24 hours per day all year long. Likewise, if the motor is much higher in horsepower rating, the savings also come back quicker. The bottom line is that a VFD will save money in a centrifugal application. How much and when the investment of the drive will be returned are answered in each individual application by hours of actual motor running, the cost per kilowatt-hour, and how good a deal one gets on that drive purchase. When the rebate is a possibility, then implementing the AC drive should be done as fast as possible; this is because the rebate period will expire and every hour the motor runs without a VFD controlling it means more energy dollars not saved!

Fan and Pump Curves

Understanding the fan curve in Figure 9–5 and what the various points on the curve mean to the operation of a fan system will allow the theory of the affinity laws to be applied. The overall fan curve *A* is typical for the fan and motor in place. It indicates that at 100 percent pressure there will be zero percent flow and at 100 percent flow there is zero pressure in the system. The natural curve *B* is also called the *system curve*. Every duct system has a natural condition about it: the length of ductwork, the bends and pressure drop through it, the diameter, and transitions in the ductwork. This all provides the data points for the natural system curve, with each point of flow having a specific pressure. The last curve shown, *C,* is the *artificial curve.* It is artificial because dampers, inlet guide vanes, or other mechanical restrictors have to be employed to get the various flows at the desired static pressures.

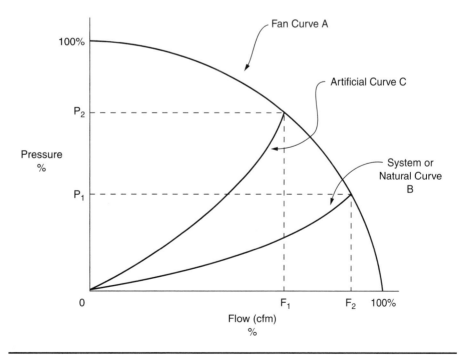

Figure 9–5 Fan curve with *no* VFD in operation.

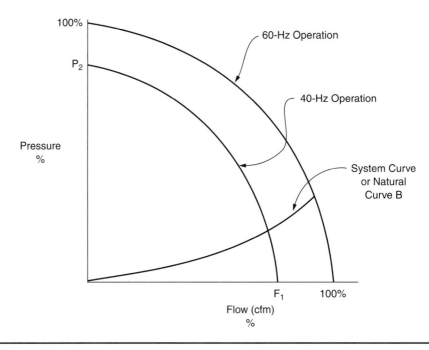

Figure 9–6 Fan curve with VFD in operation to "hit" the flow and pressure points found with dampers.

The fan curve shown in Figure 9–6 again illustrates the fan's system curve, which is also the 60-hertz operating point of the motor (full speed). However, in order to attain desired operating points *D,* the dampers could be used to get that flow and pressure or the motor speed could be turned down to 40 hertz (*D*). To turn the motor speed down, there must be a VFD in the circuit to make it possible, thus following the affinity laws and their close relationships to speed and power savings. The area above the 40-hertz curve and below the 60-hertz curve represents the energy savings attained.

Energy Policy Act

The Energy Policy Act (EPACT) of 1992 was introduced and finally enacted to minimize the United States' dependence on foreign oil for security reasons, to reduce environmental impacts of energy production, and to maintain the technological competitiveness of the United States. This act has made an impact. While there have been numerous energy-related acts and regulatory guidelines over the past 20 years, the Energy Policy Act of 1992 is the most far-reaching and the most quoted (much more so than the Clean Air Act). Because EPACT will be enforced, users and manufacturers are taking action. Electric motors have to meet certain new and higher efficiency ratings. All state and government facilities, even abroad, were to begin programs that would save that facility energy in the short- and long-term. Electronic ballasts and VFDs were to be used wherever possible. This energy act also allowed, for the first time, wholesale generators of electricity to apply to the Federal Electrical Regulatory Commission (FERC) to order the transmission facilities of other companies to grant access to send their power to other wholesale generators (Section 721). This did not apply to wholesale customers accessing transmission facilities to send electricity to retail customers, since this is typically within the jurisdiction of state regulators. However, the deregulation of the electric power companies is now a reality at the retail level.

Besides its relevancy to energy-efficient motors and VFDs, EPACT covers total energy efficiency, electric vehicles, renewable energy, and alternative fuels. The process of its implementation, culminating with the issuance of FERC Orders 888 and 889, has brought on a rapid rearranging of corporate and regulatory institutions within the industry. Congress wanted conditions necessary to create a competitive electric power market that would place greater reliance on competition and discipline prices and bring out the greatest overall economic efficiencies that utilities and industry can achieve.

Optimizing a Motor-Driven System

Not every AC induction motor in service is going to be retrofitted with a VFD. Whenever the conditions suggest, applying a VFD to an AC motor can produce significant energy savings (and these applications are many). With energy savings able to come from many areas, any motor-driven system has to be fully evaluated. If a VFD is not the choice, then that motor-driven system still can be affected to make it energy efficient.

The following is reprinted (with permission) from the U.S. Department of Energy's Office of Industrial Technologies BestPractices reference materials. Call the Information Clearinghouse at 1-800-862-2086 or visit the website, http://www.oit.doe.gov, for additional information.

Motor-driven systems can be analyzed in four distinct areas for efficiency: (1) the *power quality* of the overall system, (2) the *motor-control* method, (3) *motor and transmission efficiency,* and (4) *monitoring and maintenance* procedures. A VFD obviously falls into the *motor-control* category; however, whenever a VFD is incorporated, the power quality, motor and transmission efficiency, and monitoring and maintenance issues affect the VFD's performance somehow. A motor *and* drive system work in unison and to obtain maximum performance (i.e., maximum efficiency and use of power); it is mandatory that *all* the following factors be considered.

Power Quality

It is important to design and install electric systems that meet safety codes, minimize downtime, and reduce electric losses. A qualified electrical engineer should oversee any major electric system modifications since poor power distribution within a facility is a common cause of energy losses.

Note: Existing facilities should be checked periodically for electrical problems. Since electrical codes are designed primarily for safety, optimizing efficiency often means surpassing code requirements.

Maintain Voltage Levels   Voltage at the motor should be kept as close to the nameplate value as possible, with a maximum deviation of 5 percent. Although motors are designed to operate within 10 percent of nameplate voltage, large variations significantly reduce efficiency, power factor, and service life. When operating at less than 95 percent of design voltage, motors typically lose 2 to 4 points of efficiency, and service temperatures increase up to 20°F, greatly reducing insulation life. Running a motor above its design voltage also reduces power factor and efficiency. Because voltage decreases with distance from the step-down transformer, all voltage measurements should be taken at the motor terminal box.

Minimize Phase Unbalance   The voltage of each phase in a three-phase system should be of equal magnitude, symmetrical, and separated by 120 degrees. Phase balance should be within 1 percent to avoid derating of the motor and voiding of manufacturers' warranties.

Several factors can affect voltage balance: single-phase loads on any one phase, different cable sizing, or faulty circuits. An unbalanced system increases distribution-system losses and reduces motor efficiency.

Voltage unbalance is defined by NEMA as 100 times the maximum deviation of the line voltage from the average voltage on a three-phase system divided by the average voltage. For example, if the measured line voltages are 462, 463, and 455 volts, the average is 460 volts. The voltage unbalance is:

$$\left(\frac{460-455}{460}\right) \times 100\% = 1.1\%$$

Maintain High Power Factor   Low power factor reduces the efficiency of the electric distribution system both within and outside of the facility. Low power factor results when induction motors are operated at less than full load. Many utilities charge a penalty if power factor dips below 95 percent. Installing single capacitors or banks of capacitors either at the motor or the motor-control centers addresses this problem.

Maintain Good Power Quality   Motors are designed to be operated using power with a frequency of 60 hertz and a sinusoidal waveform. Using power with distorted waveforms will degrade motor efficiency.

Select Efficient Transformers   Install efficient and properly sized step-down transformers. Older, underloaded, or overloaded transformers are often inefficient.

Identify and Eliminate Distribution-System Losses   Regularly check for bad connections, poor grounding, and shorts to ground. Such problems are common sources of energy losses, are hazardous, and reduce system reliability. A number of specialized firms can search for such problems in the facility using electric monitoring equipment and infrared cameras.

Minimize Distribution-System Resistance   Power cables that supply motors running near full load for many hours should be oversized in new construction or during rewiring. This practice minimizes line losses and voltage drops.

Motor Controls

To reduce electric consumption, use controls to adjust motor speeds or turn off motors when appropriate. For example, equipment often can run at less than full speed or be turned off completely during part of a process cycle. When correctly used, motor controls save significant amounts of energy, reduce wear on the mechanical system, and improve performance.

Use Adjustable-Speed Drives (ASDs) or Two-Speed Motors Where Appropriate   When loads vary, ASDs or two-speed motors can reduce electric energy consumption in centrifugal pumping and fan applications—often by 50 percent or more. For more information, consult the *Adjustable-Speed Drive Application Guidebook*.

Consider Load Shedding   Use controls to turn off idling motors. NEMA Publication MG-10, *Energy Management Guide for Selection and Use of Polyphase Motors,* includes comprehensive load-shedding guidelines. The maximum number of on/off cycles per hour and the maximum amount of time off between cycles are affected by motor size, type of load (variable or fixed), and motor speed.

Motor and Transmission Efficiency

When replacing a motor, purchase the most efficient, affordable model. Motors with a wide range of efficiencies are available in most classes (horsepower, speed, and enclosure type). An energy-efficient motor will typically cost 10 percent to 20 percent more than a standard model. However, this higher cost is often repaid in less than 2 years through energy savings. Optimize motor efficiency by making certain the motor is properly sized. Oversizing and underloading can lead to low power factor and increased losses.

Choose a Replacement Before a Motor Fails   If waiting until a motor fails occurs, the primary concern will be speedy replacement. Evaluate all motors in the facility. It may be cost-effective to save energy and increase reliability by replacing some working motors with correctly sized, energy-efficient models.

Develop a replacement plan for all critical motors. Decide which motors should be replaced with an energy-efficient or smaller-sized model upon failure. Then, contact motor distributors to determine whether the desired energy-efficient motor model will be available. If not, consider purchasing critical replacement motors now as back-ups.

Choose Energy-Efficient Motors   Select the most efficient motor possible within the desired price range. An energy-efficient motor that costs up to 20 percent more than a standard model is typically cost-effective if used more than the number of annual hours listed in Table 9–1. This table lists the minimum annual hours the motor should operate depending on the cost of electricity (¢/kWh).

For example, if electricity costs 4¢/kWh and you are replacing a motor that runs as few as 3500 annual hours, an energy-efficient model would be cost-effective with a 3-year payback.

Energy-efficient motors offer other benefits. They are usually higher quality, are more reliable, have longer warranties, run quieter, and produce less waste heat than their less-efficient counterparts. Many utilities offer rebates for energy-efficient motors and other efficiency improvement.

TABLE 9–1   **Selection Criteria Table**

Minimum Payback Criteria	Average Price Per kWh			
	2¢	4¢	6¢	8¢
	Annual Hours			
2-Year	—	5250	3500	2600
3-Year	7000	3500	2300	1750
4-Year	5250	2600	1750	1300

For more information:

- Consult the *Energy-Efficient Motor-Selection Handbook.* This publication provides motor-selection guidelines plus a comprehensive review of motor field load and efficiency estimation techniques.

- Obtain *An Energy Management Guide for Motor-Driven Systems.* This publication addresses power factor correction, preventive and predictive maintenance, and troubleshooting and tuning your in-plant distribution system.

- Acquire and use the MotorMaster+ motor-analysis software program, available through the Motor Challenge Information Clearinghouse.

- Contact your electric utility conservation office.

Match Motor-Operating Speeds   The energy consumption of centrifugal pumps and fans is extremely sensitive to operating speed. For example, increasing operating speed by 2 percent can increase the power required to drive the system by 8 percent. To maintain system efficiency, it is critical to match full-load speeds when replacing pump and fan motors.

Motor manufacturers stamp "full-load rpm" ratings on motor nameplates and often publish this data in catalogs (the MotorMaster+ database also provides full-load rpm). This operating-speed rating varies by as much as 50 rpm. In general, try to select a replacement fan or pump motor with a full-load rpm rating equal to or less than that of the motor being replaced.

Size Motors for Efficiency   Size motors to run primarily in the 65 percent to 100 percent load range. Consider replacing motors running at less than 40 percent load with properly sized motors. You can address applications with occasional high-peak loads by a variety of design strategies, including ASDs for pumps and fans, reservoirs for fluids, and fly wheels for mechanical equipment.

Choose 200-Volt Motors for 208-Volt Electrical Systems   When choosing motors for a 208-volt electrical system, use a motor specifically designed for that voltage rather than a "Tri-Voltage" motor rated at 208–230/460. Tri-Voltage motors are a compromise design that run hotter and are less efficient and reliable than a 200-volt motor operating at 200 or 208 volts.

Minimize Rewind Losses   Rewinding can reduce motor efficiency and reliability. The repair-versus-replace decision is quite complicated and depends on such variables as the rewind cost, expected rewind loss, energy-efficient motor purchase price, motor size and original efficiency, load factor, annual operating hours, electricity price, availability of a utility rebate, and simple payback criteria. Following are several rewind "rules of thumb."

Always use a qualified rewind shop. Look for an ISO 9000 or Electrical Apparatus Service Association (EASA-Q) based quality assurance program, cleanliness, good record-keeping, and evidence of frequent equipment calibration. A quality rewind can maintain the original motor efficiency. However, if a motor core has been damaged or the rewind shop is careless, significant losses can occur.

Motors less than 40 horsepower in size and more than 15 years old (especially previously rewound motors) often have efficiencies significantly lower than currently available energy-efficient models. It is usually best to replace them. It is almost always best to replace non-specialty motors under 15 horsepower.

If the rewind cost exceeds 50 percent to 65 percent of a new energy-efficient motor price, buy the new motor. Increased reliability and efficiency should quickly recover the price premium.

For further reading, see the *Industrial Electrotechnology Laboratory Horsepower Bulletin*. Also refer to *How to Determine When to Repair and When to Replace a Failed Electric Motor* and *Evaluating Motor Repair Shops*, sections of the Electric Power Research Institute and Bonneville Power Administration publication, *Quality Electric Motor Repair: A Guidebook for Electrical Utilities*.

Optimize Transmission Efficiency   Transmission equipment, including shafts, belts, chains, and gears, should be properly installed and maintained. When possible, use synchronous belts or chains in place of V-belts. Helical gears are more efficient than worm gears; use worm gears only with motors under 10 horsepower.

Monitoring and Maintenance

Preventive maintenance maximizes motor reliability and efficiency. Develop a monitoring and maintenance program for all three-phase motors based on manufacturers' recommendations and standard industrial practices.

Perform Periodic Checks   Check motors often to identify potential problems. Inspections should include daily or weekly noise, vibration, and temperature checks. Approximately twice a year, test winding and winding-to-ground resistance to identify insulation problems. Periodically check bearing lubrication, shaft alignment, and belts. A variety of specialized instruments is available for monitoring purposes.

Control Temperatures   Keep motors cool because high temperatures reduce insulation life and motor reliability. Make certain motors are shaded from the sun, located in well-ventilated areas, and kept clean, since dirt acts as an insulator.

Lubricate Correctly   Lubricate motors according to manufacturers' specifications. Apply high-quality greases or oils carefully to prevent contamination by dirt or water.

Maintain Motor Records   Maintain a separate file on each motor to keep technical specifications and repair, testing, and maintenance data. Maintain time-series records of test results, such as winding resistance. This information will help in identification of motors that are likely to develop mechanical or electric problems. In addition, these records may be necessary for the proper repair of a failed motor.

Use the aforementioned twenty items to optimize the motor-driven system. For every AC motor in the facility, make a checklist from these items and treat the individual motor as a complete entity. *Maximum efficiency is not always black and white.*

CHAPTER 10

Future Variable-Frequency-Drive Technology

With the microprocessor industry taking leaps and bounds in terms of speed, throughput, and miniaturization, the VFD industry will benefit. Breaking the VFD down into two distinct entities—the control portion and the power section—most of the leaps will be in the control section. Controls for VFDs include the diagnostics, communications links, and the basic operating structure. With new microprocessor capability, the control section should change often and for the better.

New Control Platforms

More and more data can be written into drive software and firmware. More user programming capability will be provided. Drives that have 100 and more settable parameters will continue to add to that base set. As drives become configurable as volts-per-hertz type, closed loop, and sensorless vector—all out of the same box—the ability to do such will be included with the drive (whether it is employed by the user or not). This will make the drive more application friendly and will allow the user to apply it accordingly, sometimes changing the control scheme to suit the application whenever something regarding the tougher applications has been overlooked. In essence, the drive will come equipped with "bail-out" software for the user!

As more automated process control becomes the norm, drives are and will continue to be expected to link somehow with a host—either serially (at a minimum) or via some high-speed network. This means that many available protocols will have to be written and proven to work for many master controllers that connect to VFDs. This will be true industrially (as with PLC equipment) and commercially (as with building-automation systems). New control networks providing some standardization are emerging and should help to unify VFD control and network platforms in the future. Additionally, new sensor technology will continue to provide feedback and closed-loop PID control for the drives.

As costs for VFDs continue to come down, the single, universal drive box becomes more of a reality. Since a VFD actually rectifies AC power into DC voltage anyway, it has always been possible, but not practical, to operate a DC motor. The problems here have been that if a universal package is also to run AC motors, then it must have an inverter power section on board also, which is a cost adder. A DC application may not tolerate the extra cost

156

of the hardware even though the software could have the DC speed and current-regulator routines built in. But, as overall costs come down, users may like the idea of one box so that some standardization can take place in the factory. With this concept comes the other notion that the VFD can take the place of the reduced-voltage starter. Again, this is merely a cost issue, and, as starter prices and VFD prices get closer to one another, why not buy the VFD as the soft-start device? It is inherently that!

New Power Modules

The VFD's power structure is changing every year, also. More efficient power transistors are being designed. What this effectively means is that if the switching times are better controlled then electric current can be conducted through the transistor with less heat loss. This means that the transistor will require less heat sink area for dissipation, and, therefore, the actual, physical drive size becomes smaller. Fan cooling and special firing sequences will help to keep this miniaturization process ongoing; however, there will have to eventually be a physical limit as to how small the power bridge can go.

With size and efficiency considerations for the power modules evolving, there is a net effect change to the accompanying gate driver and power-supply modules. These modules, which control the firing of the transistors, are current dependent, and their physical size reflects this. As the transistors get smaller, so will the gate circuitry. This also means that these power modules will get "smarter." They will come equipped with built-in intelligence to function more independently without the VFD's main control board's influence. These are called *intelligent* power modules.

New Designs

With harmonic distortion concerns with VFDs on every engineer's mind, future drive packages will have to address these concerns. As more and more drives are added to a facility's power grid, along with many other nonlinear loads (computers, switching power supplies, etc.), the distortion levels increase. Variable-frequency drives can come equipped with input filters or reactors to help reduce their contribution, but the twelve, eighteen, and twenty-four-pulse drive systems can do away with a great deal of the distortion. As is seen in Figure 10–1, the twelve-pulse system requires a special dual secondary transformer,

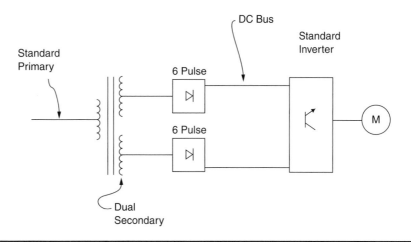

Figure 10–1   A 12-pulse input VFD.

which feeds two six-pulse diode converter bridges in the VFD. A phase shift of 90 electric degrees provides a cancellation of distortion caused by the diode rectification. An eighteen-pulse system utilizes triple-secondary on the input transformer and eighteen diodes, and the twenty-four-pulse system follows the same pattern. With diode costs coming down, this approach is being employed more often.

Harmonic concerns and high-speed communication platforms between drives and other controllers will continue to be driving forces of the technology. Local area networks (LANs) and the ability to synchronize VFDs to one another are demands that building-automation systems and factory-automation systems are making. Variable-frequency drives will continue to change in these areas.

Residential Applications

Whether or not the homeowner accepts this technology does not really matter. Manufacturers of equipment used in the home will make the change for homeowners. Washing machine makers and heating and cooling equipment manufacturers want an edge. Therefore, if their product with this new energy-saving technology helps to gain market share over their competitor, then they will find many ways to incorporate it. As costs of AC electronic drives come down, it makes it all the more justifiable when considered for use in a residential appliance. Additionally, new concepts for energy savings can be tried for their practicality.

One practical use of the VFD now in use is in the household washing machine. For years, industrial washing machines have employed VFDs on their three-phase motors. Today, the VFD is low enough in cost to be incorporated into the home's washer (see Figure 10–2). Household electricity is typically single phase, and this is not a problem to the VFD, which can accept single-phase input power and still operate and control a three-phase motor. The next feature the VFD provides is its ability to change the rotational direction of the motor. This allows the washing machine to be a front-load type, thus using less water in a cycle. This front-load configuration also means that the mechanical drivetrain is no longer cam-actuated (as is the traditional top-load, agitator-type washer), and, therefore, the washing chamber can be made to run extremely fast. The VFD allows for overspeeding of the motor to accomplish this, and, thus, the clothes get much dryer while still in the washing machine. This means that the drying time is reduced and saves electric costs over time.

The furnace fan in the home typically consumes almost twice as much energy as does the total lighting within the house. The furnace fan running 10 hours per day, with many,

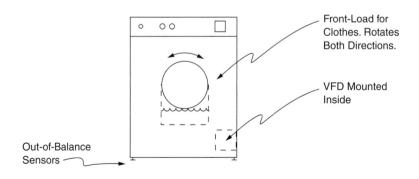

Figure 10–2 A residential washing machine utilizing a VFD.

many starts and stops, can consume over 100 kilowatt hours of electricity in a month's time. A VFD placed on this motor would help in many respects. First, any reduction in motor (fan) speed would save energy. This is proved by the fan and affinity laws of physics. Secondly, the AC motor, as it starts and is typical of starting any AC motor under load, has an electric current draw of 3 to 4 times the value usually seen while running. This inrush current does take its toll on motor windings over time. Envision 30 to 40 starts per day times 7 days per week; that is nearly 300 starts every week. Additionally, there have been some studies done to suggest that the furnace fan should run all the time. If this practice were done and the motor were run at full speed all that time, the furnace-fan costs could soar to be around 25 percent to 30 percent of the total monthly electric bill. This would not be acceptable to most, and this explains, perhaps, why this practice is not implemented. Keeping air flowing throughout the house can keep cold and hot spots from forming and, with a good filter system in place at the furnace, keep the air fresher and cleaner. However, energy costs keep this from being practiced—until a VFD is considered!

A VFD can keep a slow, steady speed at the fan motor when heating or cooling is not required. This slower speed will only use the energy needed to move a small amount of air, thus making the actual costs to operate practical. Whenever there is demand for full heat or cooling from the thermostat, the VFD can be programmed to change to a higher speed until the thermostat is satisfied and then drop the fan motor back to a slow speed. The drive allows for virtually any speed setting that the homeowner desires or needs for his or her residence and comfort level. Ironically, the energy costs associated with this mode of operation will surprisingly be less than they were starting and stopping a motor 300 times per week. The initial cost of the AC VFD and its installation shall have to be factored when justifying this application.

Another application of AC VFDs already being implemented is that of optimum, heat-pump-system control. The performance of the heat pump can be enhanced to make the system more efficient. The heat pump is not for every climate. It works in one cycle to heat a residence and reverses that cycle to cool the residence. The main drawback is that whenever the temperature outside is too low, not enough heat exchange can be made fast enough to satisfy the thermostat and eventually the occupants. This is when the high-electricity-consuming coils come on as auxiliary heat—good, quick heat but extremely expensive. The VFD cannot correct for this condition but can smooth out motor operation during the many other instances where temperatures are milder. In doing this, energy savings can be attained. A typical heat-pump system must have several circuit breakers to electrically function properly and safely. Sometimes a quantity of two 30-ampere and one 40-ampere breakers (total of three) are required for a heat-pump system. This is a possible total of 100 amperes of service that can be demanded, and this is why a residence with a heat-pump system must have a minimum of 200-ampere service to the house. There is a fan and a motor in the outside unit for air-to-air heat exchange. There is also a fan motor attached to the inside unit to move air within the house and across the coils. There has to be a small compressor motor to compress the refrigerant to move heat from place to place, and, lastly, there are those dreaded electric coils for auxiliary heat. This unit has the potential to consume a large quantity of electricity, and that is why the electric heating and cooling loads of any residence are the greatest portion of any monthly electric bill. Until the heat-pump technology is antiquated and eventually replaced, heat-pump manufacturers are continually striving to make the heat-pump system as efficient as possible. This is where the VFD can come into the picture.

Swimming pools or spas are always going to be equipped with pump motors needing control. In the case of the swimming pool, this is usually the pool's filtration system. Sometimes it is desirable only to run the filter at a slower rate rather than full filter. An AC VFD may be used in this instance. With a Jacuzzi or spa, the water jets must have the ability to spray faster or slower depending on how many people are in the spa and what the desired jet action is supposed to be. Variable-frequency drives provide exact control of these systems. Other applications of AC VFDs within the home are on some types of treadmills, whereby the speed of the tread can be adjusted by the user, and on in-house rehabilitation and water aerobic systems that allow the adjustment of the water current by the user. Thus, it can be seen that AC VFDs are quickly integrating themselves into the residential sector. Refrigeration, freezer, and compressor manufacturers are looking at this technology further to find justification to implement the drive on their equipment. As the costs of these variable-frequency controllers continue to come down, justification will be made very easy. As time goes on, more and more differing applications will be fitted with AC VFDs within the home.

Variable-frequency drives have come down so much in cost in the past 10 years that the residential marketplace can now utilize them effectively. The VFD can provide energy savings, and, if a rebate program is offered by the electric company in the region, then that makes installing one all the better of an idea. If energy savings is not a good enough reason, there is the issue of the motor's protection. A VFD will protect the motor from overloads and will keep it from burning itself up. Another attractive feature that the drive provides is soft starting. This type of starting can add years of life to the AC motor and can also save the wear and tear on the mechanical system to which the motor is attached. Another benefit! Does the application require that the speed be lowered? The best way to accomplish this with an AC induction motor is by implementing an electronic VFD. This is the quickest and smartest way to vary motor speed. Thus, there are many very good reasons to incorporate a VFD onto an AC motor these days. In the future, VFDs will be even more affordable and residential use will increase dramatically. Eventually, they will be available at the local hardware store.

The introduction of AC VFDs into the residential market has already begun. Washing machines, heat pumps, and in-house exercise pools have a variable-speed drive built into the equipment today. Homeowners and domestic homemakers will all have to educate themselves on this technology. Technology has been forcing itself upon society for years. People have had to learn different computer-operating systems over and over, have had to train themselves on different application programs for work and home, and have had to learn much more about electricity and electronics. The trend today is for manufacturers of electronic equipment to charge for support and service after an initial grace period. This can be rather expensive and not attractive to the homeowner or owner of the high-tech equipment. People will be forced to be self-trained electricians to effectively run their home's equipment.

Variable-frequency drives also lend themselves well into any home-automation scheme. If the residence is going to have many "smart" appliances and devices hosted to a master controller or computer, then the digital VFD can fit right in. The digital VFDs have several low-voltage inputs and outputs, which can be used to start and stop a motor from a remote location or computer. The desired speed can also be downloaded to the drive from the host as a voltage or current signal proportional to how fast the motor should run. Logging of fault and running data is also achieved by having a communications link from the digital VFD to the host controller. From here, the equipment can be linked to locations outside of the house via telecommunications lines. The digital AC VFD has all the features built in already.

Electric-Vehicle Applications

Variable-frequency drives on electric vehicles is today a reality and will continue to be just that. Electric vehicles, from in-plant parts movers to the family automobile, are part of an emerging technology that utilizes the inverter portion of the VFD to run, control, and operate an automobile. The electric vehicle's construction basically is an AC electric motor, DC power stored in several rechargeable batteries, and the inverter that will invert the DC into variable-frequency output to the AC motor yielding the various speeds. A machine such as this will have to be parked at night in the home's garage and have its batteries recharged. This means that the garage of the future will have to have perhaps one or two dedicated circuits within the garage for connection to the vehicle(s). Special cabling will have to be in place, and, more than likely, timers and other controls for this process will be necessary. This extra electricity draw on the home's system may also be monitored separately by the meter and utility to track usage. Obviously, large oil and gasoline producers may see sales drop dramatically from the use of electric vehicles and may strive to find a way to recoup electric energy dollars.

APPENDIX

A

English-Metric Conversions

Area Conversion Constants

One square millimeter	is	.00155 square inch.
One square centimeter	is	.155 square inch.
One square meter	is	10.76387 square feet.
One square meter	is	1.19599 square yards.
One hectare	is	2.47104 acres.
One square kilometer	is	247.104 acres.
One square kilometer	is	.3861 square mile.
One square inch	is	645.163 square millimeters.
One square inch	is	6.45163 square centimeters.
One square foot	is	.0929 square meter.
One square yard	is	.83613 square meter.
One acre	is	.40469 hectare.
One acre	is	.0040469 square kilometer.
One square mile	is	2.5899 square kilometers.

Weight Conversion Constants

One gram	is	.03527 ounce (avoirdupois).
One gram	is	.033818 fluid ounce (water).
One kilogram	is	35.27 ounces (avoirdupois).
One kilogram	is	2.20462 pounds (avoirdupois).
One metric ton (1000kg)	is	1.10231 net tons (2000 pounds).
One ounce (avoirdupois)	is	28.35 grams.
One fluid ounce (water)	is	29.57 grams.
One ounce (avoirdupois)	is	.02835 kilogram.
One pound (avoirdupois)	is	.45359 kilogram.
One net ton (2000 lbs)	is	.90719 ton (1000 kg).
10 milligrams	is	1 centigram.
10 centigrams	is	1 decigram.

10 decigrams	is	one gram.
10 grams	is	1 decagram.
10 decagrams	is	1 hectogram.
10 hectograms	is	1 kilogram.
1000 kilograms	is	1 (metric) ton.

Length Conversion Constants

One millimeter	is	.039370 inch.
One centimeter	is	.3937 inch.
One decimeter	is	3.937 inches.
One meter	is	39.370 inches.
One meter	is	1.09361 yards.
One meter	is	3.2808 feet.
One kilometer	is	3280.8 feet.
One kilometer	is	.62137 statute mile.
One inch	is	25.4001 millimeters.
One inch	is	2.54 centimeters.
One inch	is	.254 decimeter.
One inch	is	.0254 meter.
One foot	is	.30480 meter.
One yard	is	.91440 meter.
One foot	is	.0003048 kilometer.
One statute mile	is	1.60935 kilometers.
10 millimeters	is	1 centimeter.
10 centimeters	is	1 decimeter.
10 decimeters	is	1 meter.
1000 meters	is	1 kilometer.

Volume Conversion Constants

One cubic centimeter	is	.033818 fluid ounce.
One cubic centimeter	is	.061023 cubic inch.
One liter	is	61.023 cubic inches.
One liter	is	1.05668 quarts.
One liter	is	.26417 gallon.
One liter	is	.035317 cubic foot.
One cubic meter	is	264.17 gallons.
One cubic meter	is	35.317 cubic feet.
One cubic meter	is	1.308 cubic yards.
One fluid ounce	is	29.57 cubic centimeters.
One cubic inch	is	16.387 cubic centimeters.
One cubic inch	is	.016387 liter.
One quart	is	.94636 liter.
One gallon	is	3.78543 liters.
One cubic foot	is	28.316 liters.
One gallon	is	.00378543 cubic meter.
One cubic foot	is	.028316 cubic meter.
One cubic yard	is	.7645 cubic meter.

Energy and Power Conversion Constants

One joule (J) = 0.738 ft lb = 2.39 × 10⁻⁴ kcal = 6.24 × 10¹⁸ eV

One foot pound (ft lb) = 1.36 J = 1.29 × 10⁻³ Btu + (3.25 × 10⁻⁴ kcal)

One kilocalorie (kcal) = 4185 J = 3.97 Btu + 3077 ft lb

One Btu = 0.252 kcal = 778 ft lb

One watt (W) = $\dfrac{1\ J}{s}$ = $\dfrac{0.738\ ft\ lb}{s}$

One kilowatt (kW) = 1000 W = 1.34 HP

One horsepower (HP) = $\dfrac{550\ ft\ lb}{s}$ = 746 W

One refrigeration ton = 12,000 Btu/hr

One HP = $\dfrac{torque\ (ft\ lb) \times motor\ rpm}{5250}$

Torque = $\dfrac{HP \times 5250}{motor\ rpm}$

APPENDIX B

Acronyms

AC—alternating current
A–D—analog to digital
AFD—adjustable-frequency drive
AGV—automatic guided vehicle
AI—artificial intelligence
ALU—arithmetic logic unit
AM—amplitude modulated
ANSI—American National Standards Institute
AOTF—acousto-optic tunable filter
ASD—adjustable-speed drive
ASCII—American Standard Code for Information Interchange
ASIC—application-specific integrated circuit
AT—advanced technology
ATDM—asynchronous time-division multiplexing
ATG—automatic tank gauge
ATM—asynchronous transfer mode
AUI—attached unit interface
BASIC—beginner's all-purpose symbolic instruction code
BBS—bulletin board system
BCD—bit or binary code decimal
BiCMOS—bipolar complementary metal-oxide semiconductor
BIL—basic impulse level
BIOS—basic input/output system
BJT—bipolar junction transfer
BNC—bayonet nut connector
BPS—bits per second
BSC—binary synchronous communications
Btu—British Thermal Units
CAD—computer-aided drafting

CADD—computer-aided drafting and design
CAE—computer-aided engineering
CAM—computer-aided manufacturing
CAM—content addressable memory
CASE—computer-aided software engineering
CCD—charge coupled device
CDROM—compact disk read only memory
CF—carrier frequency
CFC—chlorofluorocarbon
CGA—color graphics adapter
CHEMFET—chemical field effect transistor
CIM—computer-integrated manufacturing
CIP—clean in place
CISC—complex instruction-set computer
CMOS—complementary metal oxide semiconductor
CNC—computerized numerical control
COBOL—common business-oriented language
CP/M—control program/monitor
CPI—clocks per instruction
CPU—central processing unit
CRQ—command response queue
CRT—cathode ray tube
CS—chip select
CSMA—carrier sense multiple access
CSMA/CD—carrier sense multiple access with collision detect
CSR—command status register
CT—current transformer
CTS—clear to send
D/A—digital to analog
DAS—data acquisition system
DAT—digital audio tape
DC—direct current
DCD—data carrier detect
DCE—data circuit-terminating equipment
DCS—distributed control system
DD—double density
DDE—dynamic data exchange
DES—data encryption standard
DID—direct inward dial
DIN—deutsche industrie norm
DIP—dual-in-line package
DIS—draft international standard
DLL—dynamic link library
DMA—direct memory access
DNC—direct numerical control
DOS—disk-operating system
DP—differential pressure

DPDT—double pole double throw
DPE—data parity error
DPM—digital panel meter
DRAM—dynamic random access memory
DS—double sided
DSP—digital signal processor
DSR—data set ready
DTC—data terminal controller
DTE—data terminating equipment
DTMF—dual-tone multi-frequency
DTR—data terminal ready
EBCDIC—extended binary coded decimal interchange code
ECC—error correction code
ECU—EISA configuration utility
EEPROM—electrically erasable programmable read-only memory
EGA—enhanced graphics array
EIA—electronic industries association
EISA—enhanced industry standard architecture
EMF—electromotive force
EMI—electromagnetic interference
EMS—expanded memory specification
EOF—end of file
EOL—end of line
EPROM—erasable programmable read-only memory
ESD—electrostatic discharge
ESDI—enhanced small devices interface
EXE—executive or executable
FAT—file allocation table
FBD—function block diagram
FCC—Federal Communications Commission
FDD—floppy disk drive
FDDI—fiber distributed data interference
FDM—frequency division multiplexing
FDX—full-duplex transmission
FEP—front end processor
FET—field effect transistor
FIFO—first-in first-out
FILO—first-in last-out (same as LIFO)
FLA—full-load amperage
FLC—full-load current
FM—frequency modulation
FPGA—field programming gate array
FPU—floating point unit
FRU—field-replaceable unit
FSF—free software foundation
FSK—frequency shift keying
FTP—file transfer program

FVC—flux vector control
GIGO—garbage in/garbage out
GPIB—general purpose interface bus
GUI—graphical user interface
HD—high density
HDD—hard disk drive
HDX—half-duplex transmission
HFS—hierarchical file system
HIM—human interface module
HMI—human machine interface
HP—horsepower
HVAC—heating ventilating and air conditioning
I/O—input/output
I/P—current to pressure
IBM—International Business Machines Corp.
IC—integrated circuit
ID—inside diameter
IDE—integrated device electronics
IEEE—Institute of Electrical and Electronic Engineers
IGBT—insulated gate bipolar transistor
IMP—interface message processor
IP—Internet protocol
IPC—inter-process communication
IR—infrared or current resistance (drop/comp)
IRQ—interrupt request
ISA—Instrument Society of America
ISO—International Standards Organization
JIT—just in time (manufacturing)
kVA—kilovolt-ampere
kVAR—kilovolt ampere reactive
LAN—local area network
LBA—linear block array
LCD—liquid crystal display
LCL—lower control limit
LD—ladder diagram
LED—light emitting diode
LF—line feed
LRU—least-recently used
LSB—least significant bit
LSI—large scale integration
LUN—logical unit number
LVDT—linear variation differential transformer
MAN—metropolitan area network
MAP—manufacturing applications protocol
MB/Mb—mega bytes/bits
MBR—master boot record
MCC—motor control center or metal clad cable

MCGA—multicolor graphics array
MCM—multichip module
MFM—modified frequency modulated
MG—motor generator
MHz—megahertz
MICR—magnetic ink character recognition
MIL-STD—military standard
MIPS—millions instructions per second
MIS—manufacturing information system
MISD—multiple instruction single data
MMI—man machine interface
MMU—memory management unit
MODEM—modulator/demodulator
MOPS—millions of operations per second
MOS—metal-oxide semiconductor
MOSFET—metal oxide semiconductor field effect transistor
MOV—metal oxide varistor
MRP—manufacturing resource planning
MSB—most significant bit
MSDOS—microsoft disk operating system
MSI—medium scale integration
MTBF—mean time between failures
MTTD—mean time to detect
MTTF—mean time to fail
NBS—national bureau of standards
NC—numerical control or normally closed
NEMA—national electrical manufacturers association
NFS—network file system
NMOS—negatively doped metal-oxide semiconductor
NO—normally open
NOP—no operation
NVRAM—nonvolatile random access memory
OCR—object character recognition
OD—outside diameter
ODI—open datalink interface
OEM—original equipment manufacturer
OMAC—open, modular architecture control
OS—operating system
OSF—open software foundation
OSI—open systems interconnect
P/I—pressure to current
P/O—programmed input/output
PB—proportional band
PB—push button
PBX—private branch extender
PC—personal computer or programmable controller
PCB—printed circuit board

PCI—peripheral component interconnect
PCM—pulse code modulation
PCMCIA—personal computer memory card international association
PD—positive displacement
PE—professional engineer
PE—processor element
PF—power factor
PGA—pin grid array
PIC—programmable interrupt controller
PID—proportional integral derivative loop
PIV—peak impulse voltage
PLA—programmable logic array
PLC—programmable logic controller
PLCC—plastic leaded chip carrier
PLL—phase locked loop
PM—preventative maintenance
PMOS—positively doped metal-oxide semiconductor
POST—power on self test
PPP—point-to-point protocol
PQFP—plastic quad-platpack
PROM—programmable read only memory
PT—power transmission or potential transformer
PWM—pulse width modulated
QA—quality assurance
QAM—quadrature amplitude modulation
QC—quality control
QF—quad-flatpack
RAM—random access memory
RAMDAC—random access memory digital to analog converter
RCC—routing control center
RGB—red green blue
RF—radio frequency
RFI—radio frequency interference
RH—relative humidity
RLL—run length limited
RMM—read mostly memory (same as eprom)
rms—root mean squared
RMW—read modify write
ROI—return on investment
ROM—read only memory
RPC—remote procedure call
rpm—revolutions per minute
RS232—return signal 232
RTC—real time clock
RTD—resistance temperature detector
RTS—request to send
SAM—sequential access memory

SCADA—supervisory control and data acquisition
SCR—silicone controlled rectifier
SCSI—small computer systems interface
SD—single density
SDLC—synchronous data link control
SFC—sequential function chart
SG—specific gravity
SIMD—single-instruction multiple-data
SIMM—single inline memory module
SIPP—single inline pinned package
SISD—single-instruction single-data
SMD—surface mount device
SMT—surface mount technology
SNA—system network architecture
SNR—signal to noise ratio
SPC—statistical process control
SPDT—single pole double throw
SRAM—static random access memory
SQC—statistical quality control
SQE—signal quality error
SRAM—static random access memory
SS—single sided
ST—structured text
SVC—sensorless vector control
SVGA—super video graphics array
TCP—transmission control protocol
TDM—time division multiplexing
THD—total harmonic distortion
TIA—Telecommunication Industry Association
TLB—translation-lookaside buffer
TOP—technical office protocol
TPI—tracks per inch
TTL—transistor-transistor logic
UART—universal asynchronous receiver/transmitter
UCL—upper control limit
UDFB—user defined function block
UHF—ultra high frequency
UMB—upper memory block
UPC—universal product code
UPS—uninterruptible power supply
UTP—unshielded twisted pair
UV—ultraviolet
VCR—video cassette recorder
VDM—video display monitor
VESA—Video Enhanced Standards Association
V/F—volts per hertz
VFD—variable-frequency drive

VGA—video graphics array/adapter
VHF—very high frequency
V/Hz—volts per hertz
VLB—vesa local bus
VLF—very low frequency
VLSI—very large scale integration
VME—versa module eurocard, or virtual memory executive
VRAM—video random access memory
VSD—variable speed drive
WAN—wide area network
WATS—wide area telephone service
WIP—work in progress
XGA—extended graphics array
XOR—exclusive OR

INDEX